The Cultural Work of Community Radio

The Cultural Work of Community Radio

Katie Moylan

ROWMAN &
LITTLEFIELD
INTERNATIONAL
London • New York

Published by Rowman & Littlefield International Ltd
6 Tinworth Street, London SE11 5AL
www.rowmaninternational.com

Rowman & Littlefield International Ltd.is an affiliate of Rowman & Littlefield
4501 Forbes Boulevard, Suite 200, Lanham, Maryland 20706, USA
With additional offices in Boulder, New York, Toronto (Canada), and Plymouth (UK)
www.rowman.com

British Library Cataloguing in Publication Data
A catalogue record for this book is available from the British Library

ISBN: HB 978-1-7834-8932-9
PB 978-1-7834-8933-6

Library of Congress Cataloging-in-Publication Data
Names: Moylan, Katie, author.
Title: The cultural work of community radio / Katie Moylan.
Description: London ; New York : Rowman & Littlefield International, 2019. | Includes bibliographical references and index.
Identifiers: LCCN 2018047911 (print) | LCCN 2018060313 (ebook) | ISBN 9781783489343 (electronic) | ISBN 9781783489329 (cloth : alk. paper)
Subjects: LCSH: Community radio—United States. | Community radio—United States.
Classification: LCC HE8697.95.U6 (ebook) | LCC HE8697.95.U6 M69 2019 (print) | DDC 384.540973—dc23
LC record available at https://lccn.loc.gov/2018047911

∞™ The paper used in this publication meets the minimum requirements of American National Standard for Information Sciences—Permanence of Paper for Printed Library Materials, ANSI/NISO Z39.48-1992.

Contents

Acknowledgements

This book was created in significant part from the insights of the radio and community practitioners I talked with in Arizona, Miami, Montreal, New Orleans and Toronto: Dave Ankers, Ayesha Baig, Louise Burns, Daryn Caister, Fabiola Charles, Rhonel Cinous, Jaime Cole, Rose-Philippe Coriolan, Suzanne Corley, Richard Davis, Maryse Déjean, Dr Mark-Alain Dery, Anthony Dukepoo, Colonel Matias Farias, Carline Faustin, Tamara Filyavich, Jorge Fuentes, Luis Fusté, Aldric George, Natalia Gonzalez, Blanca T. Greenwood, Justin 'Ya Favourite Lightskin' Gunderson, Thomas Humeyestewa, Terri Hutchens, Richard Ives, Curtis 'Action' Jackson, Sylvia 'DJ Sugarbear' Jackson, Joseph H. Louis Jeune, Gary Johnson, Sawssan Kaddoura, Rebecca Laratte, Theresa Lombardi, Judge Alberto Milian, Sandra E. Munoz, John-Hoa Nguyen, Daniel Ondrejka, Gabriel Otero, Jon Pressick, Wanda Quasula, Joni Sadler, Ken Stowar, Bruce Talawyma, Mark Tara, Andrew Ward, Stephanie Womble and Hector Youtsey. Thank you again for your generous and illuminating contributions.

So many thanks to Alec Badenoch, Hilary Chan, John Downing, Kathleen Eull, Anthony Gonzalez, Rachel Grahame, Alison Harvey, Carolyn Kohl, Peter Lewis, Caroline Mitchell, Chrystie Mytekiak, Maria Touri and Mary Wells for crucial conversations, critical suggestions and support. Thanks too to colleagues at the University of Leicester for support throughout the process.

My love and thanks to Sarah Moylan, Kathleen Adamson and Tom Moylan for their ongoing care and understanding. Finally, gratitude and love to Peter for multifaceted love and support at home and on the road.

The Miami and Toronto fieldwork in this research was funded by a British Academy Small Grant awarded in 2015. I gratefully acknowledge support provided by the University of Leicester College Research Fund for the New Orleans, Montreal and Arizona fieldwork and to the Research Institute for Cultural and Media Economies (CAMEo) for further assistance with the Montreal and Arizona research.

Introduction

(Re-)considering Community Radio

Community radio is often greater than the sum of its under-resourced parts. It mobilises the expressive capacities of broadcast and online radio for nuanced representation of community concerns and experiences on their own terms and in their own voices. Community radio shows and stations enable and facilitate cultural and political representation of marginalised groups when mainstream radio often fails to do so. Community radio constitutes a realised forum for debate and discussion of acute community problems, and curates and preserves local and situated art, culture, heritage, language and music. It can challenge damaging, limiting and stereotypical representations of marginalised communities. It provides a comprehensive training ground for anyone who wants to work in broadcast radio. It is often fuelled by free labour. It is analogue and digital. Critically analysing community radio crucially incorporates considering what is meant by 'community', examining cultural labor practices, identifying and unpacking complex modes of production, recognising the importance of on-air translation and exploring the aesthetics of radio. Because of the proliferation of community radio stations in the world, there is no universal format for a community radio show or station. Instead, I suggest that the comparative study of specific community radio content, production practice and structural specificities in their multiplicity can comprise a useful lens for a global scrutiny of media and culture. Considering that media forms can embody contradictory components within their productive capacities and user experiences, community radio uniquely combines what Daniel Fisher describes as 'media's peculiar entanglement of material culture (technology), communicative protocol (expressive practice), and the normative pre-suppositions such media acquire (media ideology)' (Fisher 2013: 374) and knits these components together in diverse ways—often with the joins showing, revealing the mechanics and materialities of production. In diverse manifestations of this unconscious reflexivity, the community radio form embodies a multiplicity of possible meanings produced through localised practices—in contrast to consolidated and commercialised media forms in which localised differences are more often glossed over and streamlined into a cohesive rationalising and universalising whole. In recognition of the possibilities of community sites of broadcast production, John Downing identifies the historically situated quality of the radical expressive and representative capacities of such 'media, generally small-scale and in many different forms, that express an alternative vision to hegemonic policies, priorities and perspectives', which he locates on a historicised trajectory: 'In this multifarious, seething broth that we name society, what counts as politically oppositional, as personally expressive, as experimental, as embedded in the cultural present, as heralding the public's future, as reclaiming the forgotten merits of the past?' (Downing 2001: v, vii).

In their localised scale and through an alternativism situated in grassroots production practices, community radio shows and stations can embody radical possibilities for community-led content to counter normative and reductive representations of marginalised communities circulating in the mainstream. At the same time, in its capacity for producing such diversity—of content, of production practices, of station and studio spaces—community radio reminds us of the wide and deep capacities of media form in general.

THEORISING RADIO

Working from this summary of community radio's alternative functionalities, I want to zoom out here and consider community radio first as *radio*. In particular, I want to scrutinise the ways in which community radio embodies the multifaceted functionality of the medium of radio, in its deployment of radio's capacities for producing locality and communicating a sense of collective identity. Despite ongoing proliferation of online platforms, broadcast radio remains the most widely used electronic medium in the world (Bessire and Fisher 2013). Radio's simple technologies for both production and reception continue to have crucial use value, as Lucas Bessire and Daniel Fisher argue; 'radio's rugged and inexpensive materialities have become invested with new import in the many places where topography, poverty, or politics limits access to web connectivity, computers or electricity' (Bessire and Fisher 2013: 364). In a transnational world, radio's inherent capacity for producing and reinforcing a sense of *locality* is persistently of cultural and political significance. Although radio technology from its beginnings embodied a facility for 'community-building' (Hartley 2000: 155), radio broadcasting technology has been effectively harnessed to promote national and commercial interests, usually on a massive scale. Yet radio continues to enable public participation, at least in theory (and to different degrees in different forms), as David Hendy explains,

> [Radio] employs the rhetoric of two apparently different democratic functions: one in which it claims to 'mediate' the views of the listeners on their behalf, thus carving out some form of institutionalised 'public sphere' of opinion and debate, and a second in which it claims to be an 'alternative' medium representing, not just the voice, but also the active participation of those incapable of expression through other media or public forums. In reality, of course, these functions are often conflated, and sometimes subverted altogether. (Hendy 2000: 196)

In his analysis of radio's role in reinforcing and reifying, but also scrutinising, the project of democracy, Hendy complicates definitions of participatory radio while excavating the medium's role in enabling and sustaining alternative identity formations. Radio's potential for contributing to and/or sustaining specific identity formation can be read alongside its inherent capacity for community-building: both capabilities establish radio as a crucial tool for producing and performing a localised sense of shared community. Hendy (2000) argues radio functions in three key ways within popular culture: it serves and reinforces an ideal of democratic life; it nurtures, or destroys, people's sense of belonging to communities defined by language, place or (increasingly in the developed world) patterns of consumption; and it shapes trends within popular music. Radio is perfectly placed at production and technological registers for broadcasting niche content, whether disseminated via live analogue broadcast, as podcasts, or via online streaming. At the same time, radio retains a powerful homogenising capability, realised in its ongoing deployment as the dominant mainstream medium of many developed countries. One manifestation of this is the pivotal role of mainstream public service or commercial talk radio in shaping popular opinion on key political and social issues. Hendy queries this apparent paradox, saying '[t]hese two trends raise an important question: is radio reinforcing cultural differences, or eroding them?' (Hendy 2000: 194). Recognition of this contradictory capacity is also developed by John Hartley, who identifies two competing functions of radio broadcasting in simple terms, placing its potential role as a two-way forum, recognised at its inception by Bertolt Brecht (1930), against its now-dominant function as a one-way broadcaster of significant political and social events (see also Scannell 1991). These two functions embody very different attempts at community or identity building: the one through providing an aural avenue for two-way communication across communities; the other in establishing and reinforcing national identity through a one-way transmission of national culture, located in the transmission of those same political and social events. Hartley argues that '[t]he eventual dominance of commercial broadcasting, as opposed to other cultural forms, was not inherent in radio technology' (Hartley 2000: 156). Radio technology itself, as Brecht recognised, provided at least the technical capacity for two-way communication on air (Brecht 1930), yet the development of radio content and production primarily centred on one-way transmission.

Two indicative examples of this development can be seen in short-wave radio on the one hand, historically used to broadcast internationally, serving to re-create 'locality' even to those far from home; and in transistor radio on the other hand, as it rapidly became an affordable device for information and connection in developing regions. Today, the further developments of podcasts and reliable live streaming fulfil similar functions for those with internet access. It is this inexpensive, relatively simple and evolving media technology which has become so useful, and user-friendly, for community radio stations and broadcasters across the globe. In the case studies and discussion that follow in this book, I recognise these dialectical capacities within radio for both *accessibility* (inexpensive technology; the potential for participation via phone-in shows; identification and targeting of alternative or niche communities); and for *homogenisation* (mass dissemination; mass marketing of music; potential and actual use as a propaganda vehicle; potential for enforcing biased mediation; potential for listener manipulation) and I work from this complex matrix as a point of departure to focus on ways in which radio's accessibility at expressive, social and technological registers has been mobilised across particular community shows and stations.

COMPARISONS, COMMONALITIES AND CHARACTERISTICS OF COMMUNITY RADIO

Central to community radio's importance as a representative medium is its capacity to provide alternative spaces for community-led articulation for marginalised groups under-represented in mainstream media forms. As online streaming has improved, this profoundly localised cultural form has simultaneously become global in reach and scope. Insofar as community radio's political function of facilitating grassroots representation has always been inseparable from its formal functionality (with questions of equitable representation informing station schedules and inclusive production practices), radio's function and form have both evolved alongside shifts in media platforms and provision. Today there is no singular public sphere recognisable as such; in any case, prior conceptions of a singular public sphere have always been problematic in their assumptions of that posited sphere's apparent universality (see Habermas 1989, then Fraser 1993 and Warner 2002, in particular). Instead, people avail of and engage with multiple media and cultural forms across varied platforms in everyday life, and community radio—streamed or broadcasting via analogue signal—is distinctly one of those forms. In an environment of media multiplicity, community radio content and stations are perceived less as occupying an 'other' niche audio space and more as simply comprising one option among many. While a hierarchy of production values continues to inform media use at access and interface levels, community radio stations and shows are able to sustain a greater reach than before. As broadcast radio continues to settle into online platforms for dissemination, its renewed reach across both analogue and digitised formats is generating new as well as established audiences. The growing popularity of podcasts has also generated new audiences for 'radio' formats, while nonhierarchical aggregate sites such as the Radio Garden (http://radio.garden/) have unexpectedly enjoyed substantial popularity at an international level.[1]

Defined by UNESCO as radio managed and produced by and for the community,[2] community radio responds to the diverse needs of communities in distinct and specific ways depending on local and social contexts. Therefore, production practices on the ground vary considerably in their negotiation of local factors. Thus comparative analyses of community radio content and production can enhance understanding of both the particular station or shows under scrutiny and the ways in which distinct production practices can exemplify the various possibilities of community radio. For instance, for many tribal stations on isolated reservations, the need for community-led programming is more acute and less resourced than it is for urban community programming, yet production practices and structures across these diverse local and cultural contexts have more common characteristics than differences—despite substantially different informing cultural and geographical factors. Community shows broadcasting on small local or university stations or at standalone community stations often share limited production facilities and most shows are produced, researched and presented by volunteers. Yet these shared infrastructural and labour conditions combine to produce shows and schedules unique to their locality and to the community(ies) represented. Community radio's capacity to produce deeply localised and culturally specific content, despite material limitations, means that communities are enabled to represent and express their experiences on their own terms. In this book, my exploration of community radio content, production and structures considers these cultural practices as cultural work, comprising necessary self-representation and undertaken by people with a passion for radio, who carve out time to research and broadcast their shows for free. As reflected in the title, by mobilising an assembled approach to the multifaceted object of community radio content, structures and production practices, this book seeks to recognise and excavate the equally multifaceted value of this cultural work—for both individual practitioners and those communities represented.

The idea for this comparative study of community radio first emerged during a visit to New Orleans in 2011. My first book, *Broadcasting Diversity* (2013), examines ways in which community radio encourages and facilitates ongoing self-representation by otherwise marginalised communities in Dublin, Ireland. Finding and listening to WWOZ for the first time in New Orleans, I heard an entirely different form of community radio from that in Dublin: I heard New Orleanians speaking passionately and knowledgeably about local music, connecting to local listeners through a shared love of music expressed in a richly accented collection of voices. Yet in this nuanced on-air community building, WWOZ was doing the same cultural work as the Dublin stations I had researched, but in ways particular to the station and its situation in an utterly unique New Orleanian context. When my research progressed into fieldwork in a second US city, Miami, in April 2016, my efforts to determine what defined 'community radio' became more complicated as (at that time, and now at time of writing) I could not find any stations declaring themselves as 'community stations', and yet community-facing *programmes* were easier to locate. This disconnect and difficulty persuaded me that shows and stations doing the representative work characteristic of community radio were proliferating in US contexts, negotiating cultural and linguistic diversity in their given locality in distinct and particular ways. This observation in turn cemented my idea for a comparative analysis of North American community radio. As my fieldwork expanded into three further regions, the question of 'what is community radio' became a multilayered set of enquiries. I began with the definitional determination of how 'community' provision of community radio could be determined, whether by antenna reach, license category, funding model, labour structure, programme schedule and/or programme content.

Broadcast radio in the United States historically emerged as a primarily commercial service. While a public service model was first trialled in the state of Wisconsin in 1917 (Mitchell 2016),[3] public service broadcasting provision (in the form of National Public Radio, or NPR) on a national scale only emerged much later as an adjunct to the nation's powerful commercial stations (see Mitchell 2002 for a detailed discussion of the origins of NPR). The historically inclusive approach of the BBC remit to 'educate, inform and entertain' in the UK was therefore largely absent from US conceptions of radio content and legislation. Consequently, there was little development of a community sector nor any provision made by the Federal Communications Commission (FCC) for standalone licenses. In contrast, the Canadian legislative context is closer to the structures of the UK and of many EU countries in that Canada recognises and regulates three license categories: public service, commercial and community. The Canadian legislative context expands on these categories further in several subcategories, incorporating provision for dedicated ethnic stations, Indigenous stations and campus-community stations.[4] Provision was made for licensing multiple remote Indigenous radio stations from the 1960s (Szwarc 2018)[5] and further broadcast provision for diversity of representation is enabled by the 'ethnic' category of commercial radio and the amalgamated campus-community category, in which university-based stations have a mandate to represent both students and local community groups. In the United States, some broadcast radio frequencies are better than others, or at least more powerful, and as frequency allocation is finite, powerful frequencies with greater reach are more valuable and are protected by broadcast legislation. In its oversight and regulation functions, the FCC manages broadcast radio provision and categorises station licences according to analogue frequency type. In this division, FM frequencies are more powerful and popularly perceived as mainstream whereas AM stations are considered to be more niche. To take one example, most National Public Radio stations and their affiliates, which have public service remits, broadcast on FM frequencies. In an initial move beyond this traditional division, in 2000 the FCC authorised the creation of a new, low-power radio service. As Jeffrey M. Echert explains, the 'intent was twofold: to create a class of community stations not controlled by existing media concern—that is, commercial entities—and to preserve the integrity of existing FM radio service' (Echert 2015: 240). These new low-power (LP) licenses represented a new opportunity in the United States for the creation of standalone community and nonprofit stations and were also availed of by US college stations (Fauteux 2013). Yet because of spectrum scarcity, only a few LP stations could be licensed, particularly in urban areas where the airwaves are already crowded: even LP stations take up some bandwidth. In 2010, The Local Community Radio Act further legitimated the link between LP licensing and community-facing provision and in October 2013 the FCC opened a second window for applications for new LP stations, receiving 2,800 applications. As Echert suggests, '[i]n an industry dominated by only a few voices, community stations now have the chance to raise their own voices above the din' (Echert 2015: 237).

Confusingly yet pertinently, in US regulatory discourse the word 'community' is separated from broadcast licensing guidelines and does not appear in application guidelines for LP radio licenses.[6] Consequently, the lower broadcast reach of the LP license category means that this license type is suitable for under-resourced community groups seeking to secure a broadcasting license. The discursive separation in US broadcast legislation between 'community' and license type—and therefore mandated practices—

strengthens an alternative, and often co-opted, deployment of 'community' as a station descriptor across US college radio stations and NPR affiliates in particular. Miranda Joseph (2002) argues for the romantic currency of 'community' in ways I explore further in chapter 1; her identification of its public resonance as a synonym for warm, accepting collectivity helps to explain an impetus for stations to represent themselves as 'community' stations. Additionally—or sometimes instead—self-designated community stations in the United States emphasise their nonprofit status; for example, a radio station which considers that it fulfils a community-facing remit might operate under a nonprofit designation, relying on supporters to keep the station running. Discursively, the two terms are used, if not interchangeably, certainly in ways which foreground the small and local nature of a given station's programming provision and target listenership.

Given that in the US context, a station's commitment to community representation is not formally linked to licensing requirements, in order to map the scale of such outreach activities I consider the 'community'-facing set of *commitments* of a given station as traceable by identification of the community-facing *activities* performed by a given station. A useful example for this mapping exercise can be seen in the Miami radio station WSRF, which bills itself as 'the first Haitian-owned station in the nation'.[7] Licensed for broadcast since 1980 and Haitian owned for over a decade[8], WSRF has a relatively extensive reach on the AM frequency in North Miami with a strong signal of 10,000 watts, also broadcasting in Fort Lauderdale and describing its catchment area as South Florida. The WSRF broadcast studios are located in a new standalone building in North Miami Beach, which includes station offices and meeting rooms as well as a reception area. WSRF's established presence on the airwaves, extensive broadcast reach and modern and comfortable station building all suggest the station operates on more resources than many underfunded community stations. At the same time, WSRF is committed to regular outreach activities such as weekly food drives along with other activities.[9] WSRF's North Miami Beach location is adjacent to Miami's Little Haiti neighbourhood so there is greater proximity to the station for Haitian-American community members. WSRF's programmes are largely produced and presented by volunteers, although the station also incorporates a few part-time paid workers. The schedule comprises entirely Haitian and Haitian American content, and daily shows such as *The Morning Drive*, discussed in chapter 2, have a clear mandate of community representation. In its programming, outreach activities and established Haitian American ownership of the station, WSRF fulfils the community-facing remit of a community station.

In an illustrative contrast, New Orleans station WWOZ bills itself as a 'jazz and heritage station' with a remit to 'be the worldwide voice, archive, and flag-bearer of New Orleans culture and musical heritage'.[10] Also broadcasting on a particularly strong signal but on the FM frequency, WWOZ represents itself via a polished, comprehensive website incorporating substantial content about local events and activities in addition to details of the station's schedule. While the focus of WWOZ's schedule is music, the station also actively promotes a strong sense of a New Orleanian community, reinforced through regular online and on-air provision of information about locally and culturally specific events such as the city's famous Second Line Parades.[11] A further indicative comparison can be made with the community-facing strategies and services of Native American radio stations. Generally serving one or occasionally two tribal communities, these rural stations are located on relatively remote tribal lands. Of the three LP tribal stations I visited in Arizona, two provide emergency response services. KWLP, serving the Hualapai tribe in northeast Arizona, and KUYI, serving the Hopi and Tewa tribes in northwest Arizona, each incorporates emergency response functionality so that in the event of an emergency situation on the reservation, such as flooding, fire or storms leading to a breakdown in standard forms of communication, the radio station serves as the area's communication frontline infrastructure, providing an essential information service to areas that are remote and otherwise underserved by communication infrastructures. Both KWLP and KUYI (and the third tribal station I visited, KPYT) additionally foreground community-produced programming and provide in-depth radio production training for tribal members and local high school students. In their provision of timely reservation information, broadcast training and tribal community content, these three tribal stations embody the inclusive mandate of the community radio project.

Given the diverse specificities of community radio programmes and practices, I have identified the following commonalities across the community stations that constitute the basis of my studies in this book, listed here to form a working typology of community radio characteristics:

- Relying in large part on volunteer labour (programmes/stations);
- Evincing an ongoing imperative to balance equitable community/group representation in the programming schedule (stations);
- Maintaining a fluid running order but strictly adhering to timely station ID broadcasts (programmes);
- Privileging live broadcasting *or* flexible prerecording (programmes/practitioners);
- Broadcasting from under-resourced premises, limited provision of studio space (stations);

- Operating as nonprofit entities, with a dependence on listener/public contributions/sponsorship (stations);
- Maintaining strict practitioner adherence to codes of conduct, production guidelines (stations/practitioners);
- Maintaining significant practitioner autonomy over show content (stations);
- Maintaining strict training requirement for all broadcasters (stations).

The degree to which these qualities are traceable in the programmes and stations included in this research certainly varies but each of these characteristics are present to some extent. Every radio station that I visited which defined itself as a community station shared the material commonality of limited studio spaces, although these do vary in relation to degree. Some stations have a single studio, other have two or even three, allowing some prerecording of show content or guest interviews. This typology also enables and informs my identification and mapping of some of the differences between US and Canadian community provision. Finally, a further characteristic that all community programmes and stations have in common is their inherent particularity; local community radio production practices and the resulting content is informed and infused by the specificities of their geographical locality but also by the structural materialities of the radio station.

AN ASSEMBLED APPROACH TO COMMUNITY RADIO

Given these shared characteristics across community radio practices, structures and content, I want now to outline ways in which I carry out my analysis of these community radio phenomena in order to make connections between community radio practices and their societal and political significance. With this book, I hope to explore the following questions: What purpose do community-produced programmes serve? What gaps in representation do they fill? What theoretical frameworks allow for sufficient recognition and understanding of these connections and their meaning? In their specificities and commonalities, community radio programmes and practices invite and reward multifaceted, multidisciplinary critical approaches. My analysis therefore focuses on diverse production practices and the content produced, in particular on the ways in which individual practitioners from diverse communities in the five regions discussed consider and articulate community perspectives. I examine *production contexts* (considering these processes, conflicts and difficulties); *production labour* (especially considering its volunteer nature and the politics of access and proximity to technology, opportunities and interview subjects); on-air *community representation* (comprising articulations of community identity and experience, topic selection, running order and the benefits of liveness); and the *enabling structures* that inform station guidelines, remits and policies, station materialities, resources, and the politics of scheduling. I pursue each of these avenues of enquiry so as to appropriately explore how these programmes and stations represent grassroots alternatives to normative representations of underserved and marginalised communities. This assembled set of approaches also enables me to research these intersecting community radio phenomena on their own terms, as they emerge and manifest in the diverse regional contexts of this comparative research. Through this multifaceted exploration I examine the cultural work of community radio's project of inclusive community-produced representation, variously manifested and fulfilled.

PRACTITIONER REFLECTIONS

For this analysis, I draw on on-site approaches that can, as called for by Rosalind Gill and Andy Pratt, 'engage with the specificities of different industries, workplaces and locations, and attend to the meanings that workers themselves give to their labour' (Gill and Pratt 2008: 21). Community radio, in its multiple and myriad forms, nonetheless embodies persistent commonalities of material experiences and practices, at once serving as a shared site for experiences and practices of marginalised groups yet always immersed in the specificities of individually felt elements of those experiences and practices as informed by the particularities of their identity, community, place. Thus, what emerges in my discussions with practitioners about their radio production work is a series of reflections on what it means to them to be a producer of a radio show or to produce other content for the station. If 'subjectivity is always *mediated* by the meanings which people give to their experience'—how practitioners consider and define their production work—then I believe that we can in turn consider 'the meanings cultural workers themselves give to their life and work' (Gill and Pratt 2008: 19) as forming political (because *politicised*) practice. In order to identify, consider and critically explore this work of community radio practitioners as self-defined, I arranged mutually agreed upon interviews with individuals at times and locations convenient to them. In initial conversations and emails, I established my aim to discuss their work with the given station, whether this involved producing and presenting a show, managing the station or working at the station in another capacity; for example, as a production technician. With the two exceptions of roundtable group discussions with practitioners at KUYI in Arizona and at WSRF in Miami, all other interviews were one on one. When possible, I also observed live radio broadcasts in studio when allowed to do so by the show hosts.[12]

This book is not an ethnography, although I utilise ethnographic approaches of observation (in-studio) and, in particular, in-depth interviews as conceived of by Jo Tacchi in her discussion of Ethnographic Action Research (EAR) (2015). Drawing on this conception of the in-depth interview format, I arranged detailed conversations which were guided by a few key topics but left sufficient space for organic responses, digressions and expansions (Tacchi 2015). To enable conditions for more expansive conversations, I met practitioners in places of their choosing and allotted at least an hour so that the conversation could comfortably expand. Given that most practitioners I spoke to were volunteers, we often met on their lunch hour or after their work day had ended in public places convenient for them. Other practitioners I met and interviewed at their station. All interviews with station managers and station technicians took place at the radio stations themselves. All interviews referenced in these case studies were recorded for accuracy. My foregrounding of and reliance on these in-depth interviews as increasingly central to this research emerged organically, but this approach was jump-started by the detailed, nuanced, thoughtful and therefore hugely insightful conversations I had with practitioners in New Orleans—my first fieldwork site. At this stage, in spring 2015, I was following my nose on the ground and seeing what emerged. I had been unsure of what access I might gain to community practitioners, what our conversations would be like and what they might yield for both myself and the practitioners. As it turned out, once I was based in New Orleans I was able to talk to more practitioners than I had envisaged would make time for me, and I availed of several station visits to the three stations under discussion in chapter 1. The generosity of access granted by the practitioners and management of all three stations combined with an array of reflections on everyday production practices convinced me these interviews were productive—of insights for this research, but also of practitioner reflection on their own work in most cases. Thus, in what follows, practitioner descriptions of the work of production (topic choice, decision making, reaching interviewees, listener feedback, managing community representation) are simultaneously imbued with the meaning they give this work. Our conversations usually widened to become actively reflective, as through talking about their production practice or management decisions in more detail practitioners could begin to see their work as it appeared from the outside—to a researcher like me in the first instance, but also more generally, in relation to the cultural value of such work. In this sense and by my observation, the interviews often functioned as a form of praxis, in which our conversations about what producing a community-facing show involved prompted practitioners to reflect on the meanings of their own work and to make connections via examples of practice that emerged while we talked.

The guiding focus of these conversations has been each interview's stated emphasis on individual production and station practices. However, once initially directed by this focus, in conversation we were mutually able to expand on what that work entailed as we further explored its significance for the practitioner. I consider these conversations more than interviews—not to be disingenuous but because, after initial discussion of their programme, the conversation tended to expand on the person's production work in multiple and diverse ways. I also believed, and believe now, that by focusing my guiding question topics solely around work practices I could keep the dynamic of the interview as equitable as possible. I was interviewing from an academic standpoint but I was also informed by my past experiences as a radio practitioner. This meant I could draw on my own experiences as a producer, researcher and presenter in pirate radio and later in regional radio. My knowledge of studio desk technology has certainly become attenuated, but I could easily recall planning a show, choosing topics, arranging the running order, reaching out to possible guests and scheduling on-air interviews; the range of tasks which comprise a radio show's production. My familiarity with these processes was helpful in establishing common ground but this also enabled me to better understand what these individual tasks entailed, along with the uncertainties which can accompany the production process. Each practitioner who agreed to speak to me was granting me time that was often in short supply. Most radio volunteers are busy people, made busier by doing the cultural work of radio production weekly (or daily) for free. I therefore met each person in accordance to their schedule at the best location for them so as not to inconvenience them more than necessary. In talking about work practices—mostly theirs but sometimes mine, as there was often curiosity about what I was researching this for—we could compare work-related contexts and observations. Practitioners told me more or less about the rest of their lives to different degrees and in different ways, usually depending on how aspects of their lives informed or intersected with their production practices.

This approach also meant that my primary source for determining the nature of community radio production work (so as to identify and examine its characteristics, the areas of labour which went into its production, and thus what makes it meaningful) has been the perspectives of practitioners themselves. It is therefore the case that I am accepting their own accounts, taking their word for it. Yet as I continued to conduct interviews and have conversations in different parts of North America, the above-mentioned commonalities began to surface across practitioner accounts of their experiences and their production work. Initially unforeseen in their proliferation, but in retrospect structurally and politically understandable, these unearthed structural and experiential commonalities formed the backbone to this research and to this book, which have in turn prompted me to devise more nuanced and precise ways of conceiving of the diverse and multiple value of transcultural community-led radio practices. In this way, these in-depth practitioner interviews have enabled what Tacchi describes as the project of ethnography, which she suggests has 'traditionally been used to understand different cultures in detail, and reject pre-conceived ideas and categories in favour of locally emergent ones' (Tacchi 2015: 224). Given community radio's foregrounding of local and community self-representation, at expressive and structural levels, this work needs to be further explored and considered on its own terms, in relation to its informing cultural, structural and regional context. Ethnographic approaches such as the in-depth interview facilitate research which privileges these particularities. While the selective focus on production practices rather than foregrounding participant observation or other immersive practice precludes this research from being considered as an ethnography, that field's focus on the local at everyday material and expressive levels has—I hope—enabled me to foreground the particularities of each station and informing cultural, social, and regional contexts here. Through these processes and the insights yielded, this study comprises a collective set of critical practitioner reflections on practice—and thus should be considered as a collaborative project.

More broadly, and through learning the specificities of transcultural radio production in diverse contexts and regions, in this book I consider ways in which 'subjectivity can resist, evade or exceed capitalist colonization' (Gill and Pratt 2008: 19) and ways in which the structures and created/reinforced spaces (material and symbolic) of community radio variously enable such resistance. In identifying problems with her and others' prior conceptualisations of the public sphere's capacities for accountability and critique, Nancy Fraser argues that to remain critically of use public sphere theory 'must reinterpret the meaning of the inclusiveness requirement' and discussions of the critical public sphere must foreground its two conditions of use: inclusiveness and parity, but in ways which reflect how both manifest today (Fraser 2008: 96). Fraser describes a political context typified by multiple subjectivities and experiences, more strongly resonant today than even when she was writing in 2008, as the structural damage produced by unequal representation and capacity for participation has intensified. Fraser describes the necessity of 'a communicative process in which all who are jointly subjected to the relevant governance structure(s) can participate as peers, *regardless of political citizenship*' (italics in original, 2008: 96). Yet such equitable access is, as we know, hardly guaranteed. Catherine Squires, in reference to Fraser's earlier work on the subaltern counterpublic, argues that in 'societies structured by inequalities, members of dominant groups have many advantages because they have set the spoken and unspoken rules for public speech. Even if access to public arenas is theoretically granted to all, all will not necessarily be equal within those spaces' (Squires 2002: 450). Certainly such discursive dominance pervades mainstream radio across both the public service and commercial sectors. Squires points out the active limits to and dangers of participation which can be experienced by marginalised groups: 'prevailing social norms may instill fear in citizens of marginalized publics that their ideas would at best be met with indifference, and at worst violence. Thus, the ideal of an open public sphere is difficult to realize for oppressed groups (Squires 2002: 449). Given the multiplicity of experiences and subjectivities alongside the multiplicity of media forms in use, community radio is well positioned to enable cultural recognition through enabling self-representation of marginalised groups, in particular.

RADIO POETICS AND POLITICS

To consider the multiple expressions of 'community' which comprise the work of community radio—representational strategies within the radio programme, community-led production and structural practices—I further examine and explore these expressions through the specificities of radio textual characteristics. In particular, I argue that community radio creates, performs and reinforces an everyday sense of community (and community values) through expressive modes peculiar to community radio: specifically, *accented radio* and the structures of feeling such radio produces. Grassroots production modes are read through the particularities of their cultural and regional (and legislative) contexts, enabling recognition and nuanced analysis of how community radio stations and shows variously reinforce and reproduce experienced senses of 'community'. In the process, it is additionally necessary to scrutinise what *community* itself has been understood to represent—and also the ways in which 'community' has been deployed as a concept instrumentally. Some of these articulations of community are found in and explored through practitioner perspectives, taking on Gill and Pratt's suggestion in investigating what this work sounds and feels like 'on the ground.' Generally in my radio research, I attempt to approach and analyse the radio text through a *poetics* of radio, a workable lens to enable close consideration of the aesthetics of radio. I draw on the specificities of radio textuality in my analyses to better understand how the given radio text expresses itself across multiple registers of meaning. Radio's formal insistence on a singular sensory output of sound alongside powerful mediation capacities as an everyday medium (for many if not all listeners) therefore determines that a study of radio incorporates the aesthetics of radio sound and its everyday, ordinary performativity for radio. As Bessire and Fisher suggest: 'Radio's predictable sensorium demands that its practitioners and its theorists orient perception around sound, voice, and transduction. Yet in doing so, it also requires people to objectify constantly those very senses, themselves, their worlds, and even the process of mediation anew' (Bessire and Fisher 2013: 373).

In this book, therefore, my assemblage of critical frameworks had to additionally provide for close textual analysis of the aesthetics of the community radio programme. This exploration allows for and seeks to uncover further the components of a radio show's *radioness*—that is, those qualities that typify radio content and on-air aesthetics. For me, as developed in my previous work, the concept of *accented radio* usefully encapsulates radio's aesthetic qualities (voice, tone of delivery and of discussion) as they are deployed in transcultural production, which embodies the situatedness of the producer. Building on Hamid Naficy's (2001) formulation of accented production, the concept of accented radio produced recognition of how these aesthetic qualities are shaped by the conditions and standpoint of the individual producer and simultaneously enabled by and within particular material conditions and institutional structures. Accented radio, a concept I developed specifically to identify and draw out deployments of voice and particularities of accents and ways these signify and articulate situatedness, foregrounds consideration of how the accented voice is mobilized in articulations of community identity (Moylan 2013, 2018a). Studies of community radio and community and radical media emphasise the importance of facilitating voiced delivery and 'providing an account of oneself' to use Nick Couldry's (2010) formation. Voice functions at both aesthetic and political registers in a radio text and is 'inseparable from its technological and institutional mediation' (Bessire and Fisher 2013: 371) and also mobilises and makes evident relations of power of the production context, as Bessire and Fisher further argue:

> [V]oice frequently links the social potentials of vocal address to long-standing tropes of social agency and liberal personhood. . . [V]oice gains its social life through particular technologies, linguistic ideologies, and institutional practices by which radio sound is produced. Voice here comes to matter as a relationship between a global discourse of the voice as a sign of agency and self-expression and the technological and expressive practices of producing the voice for radio. (2013: 371)

Additionally and crucially, radio is often listened to on an everyday, habitual basis, which also fundamentally informs its aesthetics and production of meaning, in that as a broadcast medium it has historically been an influential medium (in ongoing ways distinct from the standalone pleasures of film, for instance). Satya P. Mohanty argues that to understand the value of an aesthetic approach we must consider how the aesthetic works upon us: 'If the aesthetic points to a dimension of our lives that is genuine but susceptible to ideological manipulation, we need to understand how and why this is so' (Mohanty 2001: 824). For me, Mohanty's approach is useful because it explicitly enables recognition of an informing context of diversity—both in relation to the object itself and in recognition of ways in which it is shaped by the context/s of production. By recognising that aesthetic value is shaped by historical and social contexts and, therefore, the 'discussion of the objectivity of value will tend to be ethnocentric and ideological even with the best of intentions', she argues that the 'best form of inquiry into the nature of value, aesthetic or ethical, will need to be comparative and cross-cultural' (Mohanty 2001: 829). This insight paves the way for critical and comparative recognition of the aesthetic value of the community radio text, in its myriad forms. Such on-air expressions take multiple and varying forms, and such multiplicity itself requires a mode of enquiry which better allows for recognition of this diversity again as Mohanty suggests: 'acknowledgement of such contextual variety and complexity is a necessary part of all reflexive and rigorous enquiry' (2001: 831).

On-air delivery in community radio is often criticised for being 'unprofessional', as it is often less polished, less slick, than either public service or commercial radio presentation, with a tendency to let callers into the show talk on without interruption. I maintain this alternative delivery style facilitates a deeper flow of conversation, at its best enabling live on-air discussion and unpacking of issues crucial to the given community. Juan Salazar and Amalia Cordova (2008) describe a media aesthetics of community or Indigenous production, where the running order is deliberately relaxed and sometimes the joins between in-studio talk and the next song demonstrate a 'poetics of imperfection'. Building on this conception, Daniel Fisher suggests this 'imperfect' aesthetic tendency in turn can serve a political function of self-representation, in which an 'aesthetics of imperfection describes the premium placed on those relationships reproduced through media's labour, address, and organisational endeavours over and above concerns with technical or aesthetic perfection in the forms produced' (Fisher 2013: 389).

Following on from this emphasis on radio textuality is a recognition that the radio text, and indeed radio's *texture* (Tacchi 1999). is infused by and simultaneously performs a sense of intimacy. Radio's power lies in this persistent capacity for intimate address in real time. Today, with our experiences of time and space further removed, truncated, reshaped, compressed and simply relocated, radio stubbornly retains a residue of being reliably intimate and immediate. It is radio's residue of broadcasting as a live medium which initially produced such a strong sense of intimacy (Scannell 1991). 'Liveness' is a key component of the radio text; aesthetically, politically and as John Durham Peters argues, philosophically. He sees liveness as denoting the communication of power, suggesting that to broadcast 'live' means that contingency is still possible, that the energy is actual, and that a new and singular event can take place' (Durham Peters 1999: 218). Unlike the complete text of the podcast, a live broadcast is ephemeral and fleeting, already past, and therefore simultaneously reminding us that the future is (also) not fixed, because a live broadcast, or stream, tells us that we still live in real time where events have not yet unfolded and the future is uncertain. In turn, the intimacy enabled and produced by the live radio text allows for and encourages emotional responses to and engagement with what is spoken: encouraging connection to stories, shared resonance of music, reaffirmed familiarity with known people, places, events. Feelings of connectedness which emerge from such intimacy can then be mobilised and reinforced in their communality through on-air conversations, stories, even arguments that the listener can feel pleasurably pulled into. The mobilisation of radio's intimacy can draw listeners together to share a sense of community through familiar music or to empathise with shared experiences of transcultural subjectivity related on-air. This capacity for producing intimacy of community demonstrates that 'when new social relations become imaginable, aesthetic notions are sometimes revised and extended' (Mohanty 2001: 825)—which is community radio's central project of inclusivity.

I attempt here, then, to assemble these multiple critical and theoretical lenses in a multifaceted exploration of what Mark Banks terms a 'sociological aesthetics' (2017: 6). Analysis of aesthetics foregrounds the form, itself comprised of multiple overlapping elements producing components of meaning, which combined can produce an object expressing meaning(s) in emotionally and politically felt ways which resonate with the hearer. For me, a sociological layer of enquiry enables an intertwining critical analysis of the cultural practices which created the object, and the decisions, material and structural factors and individual standpoints which contribute to production and shape ongoing practices. Combined, these allow for a richer enquiry into discovering: What does this cultural object do/express, and how does it express this? What (therefore) is the nature of this cultural object? What shapes the cultural work which produces this object? What are the informing conditions of the producers and of the institutions facilitating this production?

CHAPTER OVERVIEW AND STRUCTURE

Each of the chapters in this book invokes and mines recognition of community radio's dialectical project; that is, the rotation between specificities of representation and region and commonalities of infrastructure and materialities. Given these commonalities, one way of structuring chapters could have been a systematic and thematic arrangement of critical approaches to ongoing particularities of community radio: accented production, transcultural representation, community-building and/or the nuances of cultural labour. However, my strong belief was that it was more important to meet community radio where it lived, taking it on its own terms as a community-based set of practices that are fundamentally and always produced and shaped by the particularities of the specific communities represented in the place of broadcast. Therefore, I chose to base each chapter on a particular city or region as the starting point of my analysis and exploration of community radio production practices. In this way, by situating each chapter in a particular city or region, I foreground a nuanced and specific examination of community radio's imperative to 'produce locality', in John Hartley's (2000) recognition. This baseline exploration then opens into discussion of the ways in which this located study in turn grounds and enables larger communicative strategies of community self-representation. Through this chapter structure, I develop the argument that the distinct ways in which diverse community shows and stations enable and facilitate community self-representation are informed structurally and culturally by their geographical location. Working with the particularities of each region, I examine selected culturally specific programming and in turn consider this in relation to cultural functionality: of production practices, of programme content, of station practices. Miami's provision for minority community representation is different from that in New Orleans for multiple reasons which derive from both infrastructural and cultural factors. Community programming in US contexts differs from that in Canada; yet Montreal and Toronto embody quite different approaches to community self-representation, predicated in part on divergent histories of radio broadcasting. Native American tribal production practices are fundamentally shaped by cultural particularities of each tribe but also by the remoteness of the tribal lands which house the three radio stations discussed here. These examples of how differences can begin to be mapped should help to explain the ongoing critical presence of locality in the analyses to follow—not least as found in responses by community radio practitioners themselves, each of whom speak about their production practice from this intensely regional situatedness. In turn, this chapter structure is further designed for the shows and stations under discussion to be read comparatively, as points of departure for each other, so that the particularities of practices of one locality can be read against those of another, with the comparison hopefully yielding useful recognition and identification of articulation and production strategies, created by communities which resonate with their members.

In chapter 1, 'Complexities in New Orleans Community Radio', I identify production practices across three community stations in New Orleans with very different remits. In my analysis I examine and complicate normative and romanticised understandings of ‘community’ to unpack how this fluidly circulating, frequently highjacked concept can be usefully reclaimed by the community sector. In particular, I invoke David Harvey’s critical recognition of ways in which *community* as continually reproduced through ongoing discursive practices enables recognition of how community-building production practices in community radio work to reproduce these values and strengthen a community’s sense of itself. Alongside this, Miranda Joseph’s (2002) blunt identification of how ‘community’ is too often romanticised, and how this sentimental understanding can in turn depoliticise the collective workings of community self-representation, helps to frame cultural curation of New Orleanian jazz in a community station. Finally, in introducing and immediately complicating how ‘community’ can be most usefully understood as a critical nexus and set of practices, I provide a grounding concept for use throughout the book.

In chapter 2, ‘Transcultural Radio in Miami’, I address bilingual programming for the city’s substantial Haitian communities, focusing on a morning show which combines recognisable and popular breakfast talk show radio formats with transcultural delivery strategies which enable accented articulations of Haitian American community. The three young hosts produce and present their show in a fluid combination of English and Haitian-Creole. This linguistic duality and flow enables them to reach several generations of their community at both aesthetic and linguistic registers, while reinforcing their own accountability to their community. In this chapter, I introduce and expand upon the concept of accented radio, building on Naficy’s (2001) formulation of transcultural accented production, and ways in which this can be deployed to unpack and explore specifically transcultural production and delivery practices and style. With this deep focus on accented style, I also establish the value of an aesthetic consideration of radio textuality.

In chapter 3 I shift the context to Canada, starting with Montreal, which has a rich history of community programming and community radio activism. In ‘Social Justice Practices and Programming in Montreal Campus-Community Radio’, I identify and explore collective campus-community station practices and connect these with individual practitioner experiences. Building on a structural analysis of deeply inclusive station practices, I then turn to scrutiny of the cultural production labour of activist radio and ways in which this on-air activism is facilitated by the hosting station, and closely examines the particular labour of producing an activist programme. In this chapter, I develop a critical framework for considering how community radio practice can align with, shape and structure social justice programming and anticipate a more developed analysis of the cultural labour of community radio production in the conclusion. Drawing on this specific context, I explore ways in which a mobilised alternativism is deployed in campus-community station scheduling to provide space for community programmes articulating specificities of experience as well as subjectivity.

In chapter 4, 'Producing Alternative Publics in Toronto Campus-Community Radio', I draw particularly on ideas of alternative publics as conceptualised by Michael Warner (2002) and ways in which these draw on structures of feeling to build a sense of community based on actively lived experiences, as articulated on air. In my examination of the crucial ways in which community radio can facilitate the creation of alternative publics, I build on the recognition above of the necessity for effective counterpublics and anticipate a wider exploration of how community radio is usefully situated to (re)produce critical remote radio publics in the conclusion.

With chapter 5, 'The Cultural Work of Community Radio: Specificities of Tribal Radio', I return to the US context to focus on the community-building work of three stations on tribal lands in Arizona. As with previous chapters, the particularities of the production practices and structures at the three stations under discussion here have invoked their own set of critical avenues of enquiry. In this chapter, I identify and examine ways in which tribal radio production and content are shaped by tribal particularities at each station, while recognising that the community-building work of tribal radio is more acute and even less resourced than at the urban stations comprising the focus of the previous chapters. Tribal radio production embodies the commonalities of community radio practice: unpaid cultural labour, under-resourced studio resources, a remit to represent the community—but the material elements are more pronounced. For example, tribal stations have more difficulty retaining volunteer practitioners because the stations are remote and many tribal members don't have cars. The need for tribal language preservation is acute and station provision of language programming is therefore particularly essential. In this chapter, therefore, I consider the cultural work of community radio as cultural labour, broadcast training and cultural preservation, and scrutinise ways in which these practices in turn mobilise active and meaningful community participation.

Finally, in my conclusion I draw on the preceding studies and suggest ways in which recognition of the social values embodied in inclusive community radio production practices can be harnessed to inform good practice guidelines for the cultural work of community radio: incorporating recognition of both the value and the demands of volunteer production work and the efficacy of community radio in producing usefully alternative publics for articulation of diverse—and under-represented—experiences. In drawing these findings together so as to put them into action in future, my concluding remarks hopefully serve as a directional look forward into ways in which the capacities of community radio can be further mobilised at structural levels to facilitate community radio's project in materially protected and consistent ways.

Each chapter's intense focus on locality as the point of origin but also of departure situates the community shows and stations under discussion and simultaneously provides a workshop on the page for critically deploying grounding concepts and mobilising these for an overarching, and ongoing, critical understanding and analysis of the cultural work of community radio. These concepts emerge in chapter 1's critical debunking and repurposing of *community*; chapter 2's identification and exploration of *accented radio* as an articulation and production strategy for community building from the inside; chapter 3's and 4's identification and deployment of *alternativism* in community radio programming as it informs the production of *activist radio*; chapter 4's identification and discussion of ways in which community radio can produce *alternative publics*; and finally chapter 5's exploration of how self-representation of tribal communities can enable and reproduce meaningful and empowering community *participation* through tribally specific forms of delivery and representation.

A note on the photos accompanying each chapter. I suggest that community radio's capacities to reinforce a sense of community in textured and aesthetic ways can in turn facilitate and sustain a sense of collective belonging. My fieldwork photos include images of the radio stations discussed in this book but also images of each surrounding locality. In particular, I wanted to document each station's materialities, the spaces of production. I also wanted to document ways in which each station represents or brands itself. I further wanted to capture a sense of the surrounding place, to get a sense of how local contextual factors directly or indirectly informed and shaped production processes such as music or topic choice. Consequently, when considering how to represent ways in which the community programmes and the local context(s) informed one another, I chose between the actuality of station photographs and the textured nuances of community reflected in murals, street art or slogans. For the final fieldwork chapter on tribal radio, I asked the practitioners I visited to share photos they felt best represented their station. In chapter 5, therefore, the photos were selected by tribal station KWLP. Each of the photos in this book can be read as representative of some of the cultural specificities which inform and infuse each chapter's arguments.

NOTES

1. See http://www.communityradiotoolkit.net/radio-garden/ and https://www.theatlantic.com/technology/archive/2016/12/the-map-that-lets-you-listen-to-the-radio-everywhere/510368/ (accessed 15 August 2018).

2. UNESCO, with Fraser, C. and Estrada, S. Restrepo. 2001. *Community Radio Handbook*.

3. The station first broadcast as 9XM from the University of Wisconsin–Madison and was renamed WHA in 1922, taking up the convention of call letters east of the Mississippi River beginning with *W*. See also https://current.org/2016/12/roots-of-public-broadcasting-can-be-found-in-wisconsins-vision/ (accessed 26 August 2018).

4. See https://applications.crtc.gc.ca/radio-tv-cable/eng/broadcasting-services-List?_ga=2.189115627.1182026063.1534782525-120325286.1534782525 (accessed 21 August 2018) for a list of station categories currently licensed.

5. Szwarc, J., 2018, https://crtc.gc.ca/eng/acrtc/prx/2018szwarc.htm (accessed 21 August 2018).

6. See https://www.fcc.gov/media/media-bureau-forms-page (accessed 15 August 2018). Also see Fauteux (briefly) on this invisibility of 'community' in college LP radio station documentation (2013: 139).

7. See https://wsrf.com/about/ (accessed 16 August 2018).

8. See FCC pages for WSRF historical documents, http://licensing.fcc.gov/cgi-bin/ws.exe/prod/cdbs/pubacc/prod/corrp_list.pl?Facility_id=67194 (accessed 14 August 2018); the station's Haitian ownership was confirmed to me by the station manager, Rose-Philippe Coriolan, on 19 April 2016.

9. From interview with R. Cinous, F. Charles and R. Laratte on 19 April 2016.

10. From https://www.wwoz.org/about-wwoz (accessed 16 August 2018).

11. See https://www.wwoz.org/about-wwoz and https://www.wwoz.org/calendar/community, both accessed 16 August 2018.

12. A note on terms used. To reflect the holistic labour involved in producing a community radio show, in which the volunteer practitioner is often simultaneously the producer, the researcher and the presenter (and sometimes also drives the desk), I refer to these community radio practitioners variously as producer-presenters, simply as practitioners and sometimes as show hosts, a designation used often in New Orleans in particular. I move between describing shows and programmes to avoid too much overuse of either term: thus 'shows' and 'stations' and also 'programme schedule'.

Chapter One

Complexities in New Orleans Community Radio

New Orleans is in some ways an anomalous US city. First established in 1718, New Orleans has been incorporated in its current municipal form longer than almost any other North American city. Possessing a unique historical and geographical claim to a distinct cultural identity and richness, it is regularly described (by locals and outsiders alike) as the northernmost Caribbean city, imbued with characteristics of a Caribbean identity in its diversity and emphasis on musical cultures, local festivals and vivid cuisine. Yet in other ways it is quintessentially of the United States; the established infrastructural problems negotiated by New Orleans are characteristic of those faced by other struggling US cities: declining and insufficient schools, grocery deserts, inadequate health care, neglected municipal spaces, unreliable public transport. These fundamental municipal problems have only become more pronounced following the devastation of Hurricane Katrina in 2005.

Within this rich but troubled and under-resourced context, New Orleans community radio harnesses and expands on radio's capacity to *produce locality* (Hartley 2000) in multiple yet intrinsically collective ways. The modes through which a sense of locality is (re)produced anywhere is inevitably determined by the specificities of the individual city, region and/or neighbourhood. In New Orleans locality is performed, celebrated and reinforced through inclusive and regular community events and activities such as Mardi Gras and Second Line parades. Much of everyday life in New Orleans is shaped and informed by engagement with and performances of local music—widely defined, live (in venues or the French Quarter streets) or on air. New Orleans community radio programming and practices are imbued with a sense of collectivity and of shared local cultures and resonate with the city's deep relationship with music. This chapter maps and explores the diverse ways in which three community radio stations (WHIV, WRBH, WWOZ) reinforce collective structures of community, itself here widely defined and experienced. Community radio is recognised for its capacity for 'circumventing the power imbalances inherent in mass media outlets by providing a space for the traditionally voiceless and marginalised to participate in and own the media' (Ali and Conrad 2015: 7). New Orleans community stations enable articulation of marginalised experiences from the ground up, in diverse and innovative ways.

WRBH's 'first language' programming (Langer 2005, Moylan 2013) for Haitian, Spanish and Vietnamese communities reflects to some extent the city's linguistic and cultural diversity; and the station's commitment to reading to the blind on air reinforces a collective civic spirit evident in New Orleans, in community support groups across the city and in seasonal parades such as those held by the Second Line and Mardi Gras Indians. The emphasis on playing local jazz at WWOZ mines New Orleans's historical and ongoing relationship with regional forms of jazz created by intersecting histories, cultures and currents of the city and of the surrounding Louisiana regions, and incorporates interviews and live performances in the WWOZ studios by local musicians. At low-power (LP) station WHIV, founded in December 2014, the comprehensive and nuanced provision of health advice for everyday life for those living with HIV builds on recognition of the city's significantly high population of HIV-positive people, while the station's emphasis on social justice programming and content enables critique of New Orleans's ongoing infrastructural issues.

Through diverse production practices of producing community radio, New Orleans's three community radio stations (re)produce and reinforce their constituent communities in distinct and nuanced ways. At the same time, through the particularities of their broadcast practices, each station reproduces but also reworks expressions and understandings of 'community' itself. Community is a persistent and pervasive concept with multiple connotations, which I will briefly unpack here to situate the discussion of the three stations that follows. 'Community' is a fuzzy concept because of its diverse and often divergent discursive deployments, yet its fuzziness also means it is commonly used as a catch-all descriptor. Broadly, 'community' can be used in a geographically specific sense (to refer to a specific neighbourhood 'community'); in a global sense (as a unifying discursive device: 'the international community'); as a professional designation (the 'intelligence community') and as a top-down signifier of cultural, ethnic and/or linguistic affiliations, employing the definite article to infer singularity: *the* black community, *the* feminist community (Downing 2001). The notion of 'community' is often deployed in romanticised and often celebratory narratives of self-subsistence and unity, but also of inclusivity. 'Community' remains 'one of the most motivating discourses and practices circulating in contemporary society'

(Joseph 2002: xxx), thus communities can be 'our sites of hope in a difficult world' (Joseph 2002: ix). However again, 'community' has also been bound up with narrow definitions of identity, and in this sense the concept can be deployed institutionally for exclusionary purposes. Considering 'community' through lenses of enforced unity and directed identity can therefore be limiting and problematic. In recognition of this dialectical function of community as concept, I draw on David Harvey's (2001) and Miranda Joseph's (2002) reframing of community as comprised of, and produced through, a series of *practices*. Collective cultural practices can be mobilised to reproduce and reinforce a (nonetheless fluid) communal subjectivity, which can usefully and dynamically unsettle and replace the static role of 'identity' as a determining factor in defining a given community:

> While identity is often named as the bond among community members, it is a false name in that communal participants are not identical and many of those to whom an identity is attributed do not participate in communal activities. I argue that communal subjectivity is constituted not by identity but rather through practices of production and consumption. (Joseph 2002: viii)

Harvey argues for the importance of understanding 'the processes that produce, sustain and dissolve the contingent patterns of solidarity that lie at the basis of this "thing" we call a "community"' and insists therefore that community itself be considered as 'a process of coming together not as a thing' (Harvey 2001: 192). It is this sense of community as determined and shaped by ongoing processes that I find useful in considering the diverse cultural particularities of community radio practices here.

The project and work of community radio necessarily combines the idealism of romantic conceptualizations of community with pragmatism, making community radio stations sites where the practice and (re-)production of sustained community are dialectically informed both by ideals of community (as experienced by the communities on the ground) and requirements of station infrastructure (as determined by legislative and material conditions, from the top down). For community media generally, the designation of 'community' has served to define these media in particular 'as institutions responsive to demands and priorities from below' (Downing 2001: 39). If it is the case that 'to invoke community is immediately to raise questions of belonging and of power', as Joseph argues (2002: xxiii), we need to recognise that ongoing negotiations of power and representation are also central to the community radio project, traceable through work such as maintaining the ongoing flow of resources and balancing representation of diverse groups in the schedule—to give two examples from everyday community radio practice. Each of the three stations under discussion here works with and within the networks and structures of the

Figure 1.1. WRBH radio station and studios, Garden District, New Orleans, May 2015. Photo supplied by the author.

city to maintain and preserve their existence and to reinforce—and sometimes revisit—their remit. The driving force of community radio is the need for community representation: representing the needs, problems and interests of the community on air in sustained equitable ways. I suggest that two New Orleans community stations perform this function and remit through talk programming enabling participatory representation: WRBH and WHIV. A third, WWOZ, performs its representative functions in a less immediately direct but no less effective way, through music programming showcasing the diverse musical cultures of the city.

CULTURAL SPECIFICITY, READING ALOUD AND BROADCASTING LINGUISTIC DIVERSITY: WRBH FM

WRBH was established in 1982 with a remit to provide an on-air reading service for blind listeners; the station's call letters reference the full station name of 'Reading Radio for the Blind and Print Handicapped.'[1] Broadcasting 24 hours a day on 88.3 FM, WRBH at the time of writing is the only radio station in the United States providing a reading service for the blind on the FM frequency. WRBH is located in a spacious house on Magazine Street in leafy uptown New Orleans and as of 2015 had four full-time paid staff. Station listenership extends well beyond the blind community the station targets; while station resources do not stretch to research into listener numbers (as with most community stations), recurring anecdotal evidence from conversations I had in the city indicated that WRBH's live reading of three local newspapers on air each morning was a favourite feature for local professionals who listened in their cars en route to work.[2] WRBH's reading service also includes reading aloud from brand-new bestsellers, which the station receives regularly due to their established relationships with several publishers. The station is supported by famous New Orleanians such as actor Wendell Pierce and journalist and author Cokie Roberts, who come in as studio guests on return trips to New Orleans. Many of WRBH's volunteer practitioners have been producing and presenting their programmes for several years despite the precarity of this unpaid work, suggesting a dedication not only to their shows but to the station. Programmes are prerecorded and then edited, which gives volunteers, including volunteer readers, greater flexibility to arrange their recordings for the station with other commitments. Station manager Natalia Gonzalez confirms the importance and centrality of volunteers to WRBH's ongoing broadcasts:

> Without our volunteers we have no audience. And we have no listeners. And our volunteers are the people who help make us viable. . . . And nobody complains. They're underpaid and overworked. . . .
>
> Our community is so defined by our volunteers. That is why I feel this overwhelming debt of gratitude towards them. . . . We have some volunteers who have been here since the beginning. And we have some that have been here for 15 years plus. I think that speaks volumes for the organization.[3]

Gonzalez holds weekly meetings for staff to discuss what's on the schedule and any issues which have arisen. A decision not to read Mark Twain's *Huckleberry Finn* (first published in the United States in 1885) on air in spring 2015 exemplifies the different perspectives of staff and volunteers which must be negotiated in these meetings. The decision was discussed in a staff meeting in which the unease expressed by the allocated reader, Caroline Cournier, was raised and the possibility of removing the book from the reading schedule was debated at length. Gonzalez describes the collective process involved in making this decision:

> The last thing I wanted to do is make this woman uncomfortable. Make our audience uncomfortable. Yes, it's a book that there's a lot of buzz about but is it the right kind of buzz. . . . Because I think sensitivity ought to enter into this. Because I thought if I was Caroline, I would not want to revisit that . . . And also in light of the racial tensions that are in this country right now, daily, I certainly don't want to add any fuel to that fire.
>
> I told you I have a very young staff, and sometimes their youth can be an impediment, because they're all crusaders, they're all out to save the world, and it's kind of like: We're going to read this book because it was on the banned list

> and we're going to prove a point: we can read these books. But there are other banned books.
>
> . . . It's not an issue, it's just something we collectively talk about, and I would have pulled rank on this one, because I felt that strongly about it.[4]

This decision-making process, and the consideration of diverse practitioner perspectives expressed and at play within it, is characteristic of the collective practices reaffirming community values that Harvey (2001) describes. At the everyday level at WRBH, the collective nature of this decision-making process reinforces practitioners' (staff and volunteers) sense of individual agency.

Alongside the provision of its primary service of 'reading radio' for blind listeners, WRBH also provides schedule space for linguistic community representation. The weekly schedule includes Haitian, Spanish and Vietnamese programmes, all in the first languages of their target communities. *La Voix d'Haiti* and *The Vietnamese Show* began broadcasting on WRBH as responses to crises: among the Vietnamese community following Hurricane Katrina in 2005 and after the 2010 earthquake in Haiti. *The Vietnamese Show* has been produced and presented since 2008 by John-Hoa Nguyen, who works by day as an estate agent, and is broadcast between 7 and 8 p.m. on Sundays. Nguyen acknowledges that he is well known in the Vietnamese community and that this is a source of community trust which helps him in presenting and producing the programme:

> I am lucky enough to be out there in the community quite often. I would say on the West Bank, 75% of the Vietnamese know me. And throughout the New Orleans metropolitan area I would say 50% or 60% of them know me. So my face has been recognised and my voice has been recognised. Every time there is a major function in the community they will ask me to get involved, because my voice is easy to be recognised. So even though we only have one hour we are lucky enough to get the feedback from people.
>
> Every event we do, let's say the organiser looks for me, asks me to work with them, and I do not work with them as a leader, but I work with them as a link, because I can link them to different locations, to different people, and we can get together.[5]

Broadcast on Sundays between 5 and 6 p.m., *La Voix d'Haiti* usually incorporates a half hour of talk and discussion in French Creole, followed by a half hour of Haitian music. Producer and presenter Joseph H. Louis Jeune, known by the name of and broadcasting as Hector, describes the show's remit:

> The purpose of me coming here is to engage the community, to tell them what's going on, to inform the community. . . . What is important to me is to see whether I can have my community engaged.

> . . . Even though it takes a little bit of my time, to prepare a show, to come here and get it done, you know. But this is me giving back to my community, so it's really important for me to be here. And I love what I am doing, so if I have to do it for longer, I will. Definitely.[6]

Jeune stresses the importance of the Haitian show being broadcast in the 'first language' of the target audience, and describes how this informs how the community is situated in the city:

> In my mind I'm talking to the whole community . . . I have in mind everybody when I'm talking. . . . [But] I see Haitian first. Unconsciously the show to me is just Haitian, before New Orleanian. And at some point in the show I'm going to relate to New Orleans. . . . There are some resemblances between the ways we cope with these events, to make a comparison, you know. But I saw Haitian first before I saw New Orleans.[7]

'First language' programming is immediately more accessible to ethnic and cultural communities in which the 'first language' of a majority of members is not English but that of their home or sending country. Hearing their first language spoken on air can confer a sense of belonging in community members in advance of the content of the programme that follows. Nguyen too insists on the importance of broadcasting in his community's first language, saying that for the first generation of migrants from Vietnam,

> their first language is still Vietnamese. But for most of the second generation who were born in the US, their primary language is English. . . . My show is in Vietnamese because most of my listeners are of the first or one and a half generation. . . . Maybe now and then we have speakers in English but then we translate simultaneously.[8]

Both Nguyen and Jeune acknowledge and appreciate the institutional support provided by WRBH for their programmes. Jeune expands on the benefits of this support:

> Being with WRBH helps us a lot because we don't have that big budget . . . it was their way to participate in the recovery process, because that's what they do, you know? We've been here for five years and we never give them a dime, we are on the schedule, you know? So it's something we appreciate a lot, the community, really appreciate the partnership with WRBH a lot. So we really are not focusing on going out there, getting advertising, which would break down the show another piece again . . . because getting another 10 minutes for advertising would really throw us back, and thanks to WRBH we don't have to do that.[9]

The institutional and material support provided by WRBH crucially facilitates the ongoing production of all three 'first language' shows. Alongside their accessibility in relation to first language content and coverage of community issues is the more abstract but nonetheless crucial function of community-building in a wider sense. Nguyen discusses the importance of *The Vietnamese Show* in reinforcing a shared sense of community for New Orleans Vietnamese community members:

> The only thing that the Vietnamese have, that other people do not have, or at least we have it stronger, is we have a sense of belonging. See, once you have a sense of belonging, you belong to something, then that is part of you. If you have a sense of belonging to New Orleans, so New Orleans is your home, so we had no choice but to return. In our subdivisions, in any neighbourhoods, people live there but they don't belong to the neighbourhood, because they have no sense of belonging there. So the Vietnamese, even though New Orleans is not their primary home of their choice, because their main home was in Vietnam but they were forced to exile, so they made New Orleans their permanent home. And they feel that New Orleans is their home so they have a stronger sense of belonging. . . .
>
> The community must be a community, and you know that for us to be a community every one of us has to have that sense of belonging. And we have to instil that sense of belonging into people. Otherwise a community is nowhere to be found.[10]

This encapsulates the ways in which the Vietnamese community consider New Orleans home even as they recognise how this layers on top of their migration experiences (for the first generation) and their held sense of Vietnam as the 'original' home country. This complex identity position and sensibility is articulated and reinforced through *The Vietnamese Show* and *La Voix d'Haiti* in content, news and information about community events in the city.

CURATING MUSIC COMMUNITIES AND NEW ORLEANIAN CULTURE: WWOZ FM

WWOZ FM began broadcasting in 1980 as part of a wider project of community radio activism which created stations in several cities of the southern United States during this time period. In 1987, the nonprofit group 'Friends of WWOZ', itself part of the New Orleans Jazz and Heritage Foundation, became the financial and operational guarantor of WWOZ. Thirty years later, WWOZ is a cultural mainstay of the New Orleans media sphere, recognised and lauded for its ongoing role in promoting particularly New Orleanian forms of music, especially New Orleans jazz but also Cajun, Latin and Zydeco, and

Figure 1.2. Inside WWOZ radio studios, French Quarter, New Orleans, May 2015. Photo supplied by the author.

for playing newer artists alongside the revered stalwarts of the city's music scene. Broadcasting on 90.7 FM, WWOZ is credited for introducing new listeners to local forms of jazz as well as other established musical forms; the station's Saturday night *Soul Power* show playing funk classics is one of the station's most popular shows. At the same time, while in 2015 WWOZ's musical diversity was carefully curated to foreground the city's rich musical heritage, newer New Orleanian music genres such as bounce were generally absent from the schedule. WWOZ implements a music policy across its schedule, with guidelines and some restrictions over what kind of music is played when. As a community station which is primarily, almost exclusively music based rather than talk based, it can be argued that the work of negotiating and representing communities at WWOZ is particularly complex as the station seeks to balance multiple musical genres—each locally specific and resonant—across its schedule. Other anomalous elements of WWOZ, when compared to most under-resourced community radio stations, include the station's especially wide-reaching signal, which broadcasts station content as far as Louisiana's state capital of Baton Rouge, an hour's drive from New Orleans. However, perhaps the most significant factor in WWOZ's uniqueness but also in its functionality—and viability—as a community station is its relationship with the New Orleans Jazz and Heritage Foundation. The support provided by the New Orleans Jazz and Heritage Foundation's effective ownership of WWOZ confers greater relative fiscal security on WWOZ but perhaps also greater constraints. WWOZ defines itself as 'the New Orleans Jazz and Heritage Station, a community radio station currently operating out of the French Quarter in New Orleans'.[11] Preservation of New Orleanian musical history is a guiding principle of WWOZ activity and is incorporated into the station's mission statement, which reads 'WWOZ's mission is to be the worldwide voice, archive and flag-bearer of New Orleans culture and musical heritage'.[12] The station's project manager, Dave Ankers, describes WWOZ as 'offering a window into New Orleans' for listeners outside the city.

The station has its own substantial archive of material (recorded and digitised) held on the ground floor of a separate standalone period building in the French Quarter, five minutes' walk from the station premises on the Mississippi River. Given this concentration of historical material, it can be argued that the station negotiates a balance between *preservation* and *curation*. Preservation in this context describes maintaining an archive of historic audio and video material; in contrast, the activity of curation incorporates a layer of selection and editing for the purposes of exhibition or dissemination. The WWOZ archive incorporates a collection of past programmes, live shows and recorded interviews with founder Jerry Brock and the station's previous general manager David Freedman. A selection of previously broadcasted material subsequently disseminated on the wwoz.org website functions as a further form of online curation, as the selection process of placing this material online enables the reproduction of moments of New Orleanian music and culture for wider consumption. In addition, each April WWOZ broadcasts live performances from the New Orleans Jazz Fest, and these broadcasts are syndicated and also broadcast by other music stations from elsewhere in the United States. These live broadcasts offer listeners the opportunity to hear established and emerging jazz bands and musicians and also offer brief insights into a New Orleans everyday. For example, the Jazz Fest broadcasts from spring 2015 reference the city's then-recent adoption of a smoking ban and the repercussions of this for live music venues.

WWOZ's remit of preservation embodies a complex and deeply intertwined relationship between the nonprofit practices of the station and the overarching project of the Jazz and Heritage Foundation, and consequently the station represents a significant site of cultural representation in New Orleans. As noted, through its relationship with the Jazz and Heritage Foundation, WWOZ has greater resources than many community stations, such as additional studio space to host live performances for live broadcast. The station's resources also provided sufficient funding in 2015 to pay twenty staff and also enable production of high quality, even iconic, promotional materials: reflecting the station's popularity, these colourful WWOZ banners can be seen outside businesses and homes across New Orleans. The relationship between the foundation and the station confers substantial fiscal security to the continuing operation of WWOZ, underpinning the station's equally substantial cultural capital in the city. This cultural capital, alongside the online provision of local music—including live broadcasts from Jazz Fest—has drawn a substantial following of US and international listeners online, via the WWOZ website which has incorporated good quality streaming since 2014. Yet this careful curation of content for listeners outside New Orleans, arising from the preservation remit encouraged by the foundation, complicates what Harvey identifies as a 'dialectic between a free flow of processes involved in imagining and building something called "community" and the stolid permanence of an institutionalised political presence'; in this case, the permanence of the foundation's support but also its presence is tangible and necessary given the resources it provides to the station. In this sense, the foundation has incorporated the 'community' of WWOZ practitioners and listeners into its institutionalising logic. Within this dynamic, Harvey argues that '[t]he danger then exists that the institutionalised thing we call community will stifle the living processes that gave birth to it' (Harvey 2001: 193).

WWOZ therefore negotiates a particularly nuanced dialectical tension between what amounts to a mandate of cultural preservation and a wider responsibility felt by individual broadcasters and staff to support and reinforce a sense of New Orleanian community bonded and bounded by music. WWOZ, like WRBH, is considered with affection by many New Orleanians, and (as with WRBH) has generated resonant origin stories and stories about its return to the airwaves following Hurricane Katrina. These narratives, so bound up with New Orleans's recent history, also contribute to and reinforce a sense of community, described by Joseph as 'the authorising and naturalising origin stories on which community formation so often depends' (2002: xxv)—which works to combat the institutionalising tendencies produced by foundation ownership.

Station operations manager Jorge Fuentes describes the station's mandate as one of grassroots curation of authentic New Orleanian music:

> New Orleans being ground zero for the autonomous, for the authentic music of the United States, it's almost like a mandate to be able to identify, curate, respect it and pass it on to the local community first and then to the world at large.

Fuentes calls this 'an organic mandate' and expands on the cultural work of WWOZ, suggesting that

> if 'OZ had not become I'm sure there would have been another effort, somewhere in the city, to create a radio station to maintain and further the history and heritage that we have. I mean it's [the city's musical heritage] way too rich not to do that.[13]

Fuentes connects the station's emphasis on local music to its original remit of community representation:

> [I]n our particular case it was decided from the very beginning that it had to be about music because the music represented, the music was so deeply rooted in everything else, in every other aspect of the daily life of the city, that it made no sense to try to cover other things that a station like WWNO [NPR station for New Orleans], for example, was already covering.
>
> So this is why we don't do news, we don't do talk, we don't do sports, we don't do anything but the music, and I think folks have understood that . . . So it's almost mandated, that a station like ours has to by default represent that music, preserve that music, to continue to bring it to the future generations.[14]

By broadcasting New Orleanian jazz in particular, WWOZ reinforces not only the cultural capital attached to this established jazz form, but its characteristics as shaped by the city. Georgina Born (2005) identifies how different musical forms and formats are experienced ontologically, exploring how a jazz assemblage is bound up in process when playing together, and so does not impose a hierarchy in performance. Due to its collective and improvisational qualities, a jazz song or performance is perhaps never complete or final—simply captured in recording. This captured object of the jazz performance can in turn be (via radio): 'disseminated and known beyond its original time and location—in which form it becomes the aural means of educating and socialising other musicians and later generations, who are thereby empowered to create something new or to cover, re-work or transform the original' (Born 2005: 27).

This 'aural means of educating' other musicians has particular relevance and resonance in New Orleans, where music practice and performance infuses so much of the city's everyday life—musicians discuss scheduling gigs and instrument swaps in shops and out walking on the street by my anecdotal observation. Beyond this, jazz itself can be considered as a peculiarly New Orleanian musical form where the emphasis is on a live performance which functions as 'a dialogical participatory creative act grounded in an aesthetics of collaborative improvisation, one that entails a particular experience of musical intersubjectivity and place, in which the interaction is at once musical and social' (Born 2005: 27). The emphasis on collective (and social) production and performance, on improvisation and on intersecting styles coalesce into a particularly New Orleanian sensibility—as well as its own musical form. Building on this, WWOZ's showcasing of such a form can certainly be read as preservation of past musical history on the one hand, and cultural recognition of New Orleanian habits of (social) production on the other. Certainly, alongside WWOZ's preservation impetus, everyday production practice at the station is characterised by the fluidity of community radio practice typified by a more open running order, leading in turn to a flow of music and/or talk less subject to interruption. In contrast to WRBH's production practice of prerecording most shows, WWOZ's programmes are usually broadcast live from their studios, as Jorge Fuentes explains:

> One of the definitions of radio is that we live and die by the clock. Right? We have a very fluid clock at 'OZ . . . But the top of the hour ID, we need to stick to that. However, everything else is very liquid at the station and that's what I think people appreciate. You know there are things that we do . . . the encouragement, and sometimes . . . the fact that we have a hands-off approach to the DJs, we don't really tell them programmatically what music to play, that also overflows, we don't tell them how to sound or what to say, we let that overflow.[15]

WWOZ's community representation remit is fulfilled in part through the station's commitment to representing the diversity of New Orleans's myriad accents. As with John-Hoa Nguyen's vision for *The Vietnamese Show* and Hector Jeune's description of his community as 'Haitian first', this sounded sense of a community on WWOZ reinforces a collective sense of the community(ies) represented, and addressed, by the station. Jorge Fuentes feels this sense of community is communicated by *cadence*, which he expands on drawing on his own situatedness as a Mexican American:

> I love the concept of cadence. We speak differently depending on what part of the city we are in; we have different influences from all over the world. You can hear it in my particular voice that, you know, had I not told you, you already knew that I was not born here, but you also notice, I think, that in some of the things I say in English, I have become a New Orleanian and not necessarily a standard American.[16]

WWOZ station manager Dave Ankers (who became WWOZ's content director by time of writing) echoes this emphasis on producing a heard authenticity in station content, which he notes is actively encouraged by the WWOZ board; however, he also points out that for him this in itself is not the main focus or remit of the station:

> There's a push to have an authentic New Orleanian thing. Which is not necessarily the core mission. The core mission is that it's volunteers, and the volunteers represent everybody; the second level of the mission is that it's playing the music of the community.[17]

Ankers expands on this and problematizes the station remit of preservation in the context of a rapidly gentrifying New Orleans, in which the city's established cultural capital has perhaps taken on an added shine for newly arrived cultural migrants:

> I think that New Orleans is a city in transition, and that things that have been set in motion in the rest of the country have been accelerated here. There's waves of people from somewhere else coming in. And then there's issues about cultural appropriation, something that 'OZ comes in on in a fairly conservative way. Saying that we're keeping the culture; we're trying to protect and preserve this culture as it is. So the nature of the station should be a little conservative . . . but there's room for all the other music that's happening here, people come in and are doing different things.[18]

Perhaps against any commodifying tendencies of such cultural appropriation, Fuentes summarises WWOZ's accessibility and resonance for the communities it serves thusly:

> What I like to do is capture real things. And radio can be a conduit for doing things, by the sheer nature of radio being an entertainment medium, you can actually be synthesising something. But we don't, we try not to, we try to leave it as raw as it is: I want you to say y'all if that's how you speak. Because we want to reflect that. We don't want to reflect anything that it's not. And it doesn't matter if it's a New Orleans thing or not. That's the whole point. I think it's a larger situation, it's a way of saying we'll take you for who you are. You love the station and want to help us by reading out a spot, knock it out. Knock yourself out, do it because we'll love you the way you do it.
>
> I think the beauty of the way we do things is, we recognise these things, we embrace them and then we just make them work for us. And I think it works. The station sounds real. That's one of the things people keep telling us: you guys sound so real.[19]

Alongside Fuentes's perspective of ways in which WWOZ encourages inclusivity and acceptance, Ankers identifies the centrality of everyday cultural productivity for New Orleanians, which he suggests is also reinforced in station content:

> Being involved in any culture is a hobby to most people. In New Orleans it's different, it's more of a lifestyle. Everyone here is disproportionally culturally active. . . . It's a way of life here, for everybody, whether you're a newcomer or a native, or whatever you might be, of whatever socioeconomic background. It's accessible to everybody. It's a very democratic thing.[20]

Drawing on these practitioner perspectives, I suggest that the textural coalescing of regional spoken accents broadcast at WWOZ, alongside carefully selected and programmed music and the provision of culturally important local information—from live music event listings to reading the routes for upcoming Second Line parades—combine to produce a 'structure of feeling' which functions to (re)produce a collective sense of community. Structures of feeling, as developed conceptually by Raymond Williams, comprise a set of 'meanings and values as they are actively lived and felt', producing fluid and historically variable 'relations between these and formal or systematic beliefs' (Williams 1977: 132). Over time these articulations accrue to produce a structure of feeling which allows for and enables nuanced and contradictory expressions of community: cultural, musical or otherwise felt and experienced.

Part of this produced structure of feeling is the articulation of collective, potentially cathartic activities in the city. Curtis 'Action' Jackson compiles the schedule and routes of New Orleans's multiple Second Line parades for the WWOZ website. He is well placed to do this as he has established relationships with all twenty-nine of the city's social clubs which each put on a Second Line parade each year. Jackson has been working at WWOZ for eight years, and also regularly contributes to a nightly show on commercial New Orleans station Q93. He was approached by WWOZ to cover Second Line listings for the WWOZ website and is from the Lower Ninth Ward originally. He describes the significance of covering the Second Line parades:

> WWOZ goes nationwide, goes around the world, and the purpose of me being involved in this culture, and one of the purposes of me talking about it is I'm trying to expose this culture to the world.
>
> . . . I help them to get exposure, for them to come and show off, show their outfits. But it's way bigger than that. Everybody concentrates on the Second Line itself, but throughout the whole year this groups, they have fundraisers, they have giveaways, they feed the homeless, they give away back to school supplies, it's just so tremendous what they're doing and everybody's concentrating to come to the big party on Sunday. And there's way more than that. So I try my best to bring them up here, there's twenty-nine clubs, I call the presidents, I bring the presidents up to 'OZ, and then we do a podcast interview for 20–30 minutes.[21]

These podcasts are then published on the WWOZ website. Through this regular provision of information, Jackson updates New Orleans residents on Second Line activities and events—including but not limited to the Second Line parades—and also provides an aspect of New Orleans culture to website visitors outside New Orleans. Further, Jackson provides concrete and useful information for community members:

> This is just to help everybody who do come to the Second Line on Sunday—which I hope they are doing for this Sunday—or if they have one of these fundraisers or one of these giveaways, to let the public know. Like your kid's going back to school, go to this park, they're giving away free school supplies, book sacks, pencils, everything that you need to go to school.[22]

However, for Jackson, the Second Line parades fulfil a deeper, more fundamental purpose enabling a cathartic and collective celebration, which he recognises as a necessary community function:

> It's like the Mardi Gras Indians; it's about being pretty, it's about having on pretty stuff. But every group don't concentrate on expensive clothes and shoes. They concentrate on culture. It's like four hours of therapy, for everybody. That's my definition of a Second Line: it's four hours of therapy, because it contains two of your major factors that you need in life: and one's dance, and the other's laughter. If you can dance and you can laugh, it kinda eases the pain of whatever you're going through. You might have a situation this Sunday and you have a light bill due Monday, and you can't think about paying it right now but you know you gotta pay it, but I'm not going to worry about that for these four hours. You gonna pay your light bill tomorrow, but right now you're just going to go out there just to relax, you bring your kids, you bring your family, we're dancing, we're laughing—four hours of therapy.[23]

The inclusion of Second Line information and the provision of live music listings, updated daily in accents culturally specific to New Orleans, work to reproduce particularities of the musical and cultural communities the station represents. At the same time, these aural articulations confer an authenticity which draws in listeners outside the city, and in doing so exemplify, I suggest, what Miranda Joseph calls 'the performativity of non-profit production' (Joseph 2002: 117). Joseph complicates positive assumptions about nonprofits, identifying, in certain contexts, the ways in which nonprofits can function to reinforce capitalist logics of subject production:

> The promotion of non-profit nongovernmental organisation as instruments of hegemony-building depends on the notion that, despite the fact that these organisations are actively and instrumentally encouraged, participation in such organisations is voluntary, is an authentic expression of real 'grassroots' communities and their interests. (Joseph 2002: 111)

I would substitute 'assumption' for 'fact' here; however, Joseph's point that voluntary participation therefore confers an authenticity of expression is useful for unpacking the intricacies of WWOZ's operational instrumentality alongside its continuing cultural value in representing and (re)producing

Figure 1.3. WWOZ banner, French Quarter, New Orleans, May 2015. Photo supplied by the author.

communities of the city, and in its well-resourced and important project of archival preservation. Ultimately, while WWOZ disseminates elements of its culture online for wider consumption, it remains a local station. In discussing WWOZ's remit, Fuentes recognises its established nature for many New Orleanians: 'We're not proselytizing to our own neighbours. Because we expect them to take it into account that they know that we exist because they know that Second Lines exist, that Mardi Gras exists and that band practices exist'.[24]

ACTIVISM ON AIR: WHIV FM

WHIV FM was awarded a low-power (100 watt) license in 2014 to expand the outreach work of the nonprofit New Orleans Society for Infectious Disease Awareness (NOSIDA), by raising awareness about preventing and living with HIV through broadcast content.[25] WHIV has taken its outreach remit and expanded on it to provide an array of programming under an inclusive social justice remit. The schedule combines syndicated shows such as *Democracy Now* and the podcast series *The Abortion Diaries* alongside programming representing an array of New Orleans communities, interests and groups.[26] Locally produced shows broadcasting weekly cover topics from cycling (*Out Spok'n*, Tuesdays 7–8 p.m.); everyday life with HIV (*Proof Positive*, Tuesdays 6–7 p.m.), specialised cooking (*Culinary Shamanism*, Tuesdays 1–2 p.m.), managing substance recovery (*Rockin' Robyn's Road to Recovery*, Thursdays, 4–5 p.m.) and shows featuring a range of diverse musical genres, ranging from gospel to Haitian to underground to protest songs. The call sign overtly references the station's focus on HIV-positive listeners in a deliberate move to remove the stigma of the acronym. Station founder Dr Mark-Alain Dery and the station manager in 2015, Andrew Ward, point out on the station website and in promotional materials that the letters simultaneously spell out 'honouring independent voices',[27] which Ward sees as equally central to their remit.[28] When initially putting together the schedule for WHIV in early 2015, Dery and Ward scheduled the talk show *NOLA Matters* for 5–6 p.m. Monday through Friday to discuss infrastructural, social and political issues in New Orleans from a different infrastructural perspective each day. Thus, Monday's *NOLA Matters* show, 'Health Is a Human Right' is presented by Dr Dery, an infectious disease specialist who founded NOSIDA as well as WHIV. Dery uses this broadcasting hour to discuss health matters in particular. Tuesday's *NOLA Matters*, 'End All Wars', focuses on coverage of human rights and social justice issues in the city and Louisiana; and Wednesday's show, Civil Rights with Judge Calvin Johnson, foregrounds coverage of civil rights issues from a legal perspective. Thursday's show, 'Islam in the Crescent City', focuses on Muslim perspectives and experiences in New Orleans; and Friday's show, 'The Public Health Radio Hour with David Roston', returns the focus back to health matters.

WHIV first went on air on World AIDS Day, 1 December 2014, and began broadcasting regularly in spring 2015 as the schedule developed and Dery and Ward recruited volunteer producer-presenters. In foregrounding talk radio programming across the schedule, which functions overtly and directly as a community forum for local issues, WHIV's social justice mandate is reflected in inclusive programming. In 2015, programmes included *The Kickback*, focusing on young people's issues; *La Cuidad Escondita* (translated as the hidden city) which foregrounds Spanish-language coverage of issues negotiated by New Orleans's newer Spanish-speaking migrant communities; *Proof Positive*, a multifaceted show combining 'truths about HIV' with explorations of various issues; and *Expanding the Rainbow*, targeting New Orleans's LGBTQ communities and allies. When asked in an interview how he envisioned WHIV would be representing local community groups, Dery pointed to the programme schedule on the station office wall, listing programmes dedicated to different community groups (defined variously by age, health status and sexual identity among other identity positions), saying 'this is community right here'. He expands on his and Ward's plans for the station:

> This is our community. This is our New Orleans community. Voice the voiceless. This is what I am trying to do for people . . . It's 'we', it's really 'we'. What we are trying to do is create a face of New Orleans that doesn't exist. . . . [T]he other great thing about being a radio station today, is that you get to come out punching and swinging right off the bat and not be afraid to speak truth to power.[29]

A focus on production practices and the discursive work of these programmes enables a reading of community as including 'all who participate in the constitutive battles whether with acceptance of or resistance to the implied parameters and duties of communal membership' (Joseph 2002: xxiv); or, inclusion by action and participation. WHIV's stated emphasis on social justice programming enables and facilitates an on-air activism articulated through broadcast debate on current New Orleans issues alongside provision of key information for marginalised groups in the city by experts in the fields of health, housing, law, migration and other areas associated with infrastructure. Writing about queer publics and spaces of activism, Michael Warner suggests that we consider 'a concept of activism as an informal mode of representation' (Warner 2002: 210). Community radio capacities for on-air activism are explored more in chapter 3's discussion of activist programming in Montreal. However, WHIV FM provides an in-depth, nuanced and multifaceted example of an activist station enabling representation of diverse affiliated and allied groups in the city. As of 2018, WHIV's central slogan on the station website reads: 'We are not a radio station with a mission. We are a mission with a radio station',[30] reaffirming their commitment to a focus on on-air activism. At the same time, WHIV comprises the third pillar of New Orleans's community radio provision for the city's diverse, overlapping communities of interest and experience, a multiplicity expressed and engaged with in different ways by each of the three stations.

CONCLUSION: COMPLICATING 'COMMUNITY'

'Community' can be understood as a series of reflective practices through which the given community is defined and reproduced by discursive struggle (Harvey 2001). Such struggles over the limits and range of a given community work to then shape that community, its sense of itself and the 'identities of those within its sphere of influence' (Harvey 2001: 192). If community radio stations are to function as meaningful agents and sites of social change, Harvey argues that 'such movements must remain strongly nurtured by continuous processes of solidarity formation and reaffirmation.' Thus 'one of the prime means whereby a community can remain alive to its constituents and resist the deadening effect of becoming "for itself" is to be embedded in broader processes of social change' (Harvey 2001: 193). I suggest that in its curatorial role, WWOZ is becoming its own brand 'for itself'. In contrast, in its project of actively negotiating overlapping and sometimes contradictory community interests, needs and groups, WHIV is firmly and consciously situated in a project promoting social change.

Finally, while firmly and fundamentally embedded in their local communities, at the same time all three stations transcend the boundaries of their immediate surroundings and perform translocal functions from their New Orleanian situatedness, albeit in distinctly diverse ways. As seen above, while there have long been multiple ways of conceptualising and defining 'community', with the increasingly important role of online communities defined by shared identity and interest rather than geography, boundaries and understandings of what community means are becoming ever more fluid. At the same time, as online streaming of community radio has become widely accessible due to improved technology and streaming speed capacities, community radio's capacity to transcend the place of its originating production continues to expand. Yet community radio's content and practices remain culturally situated in the geographic locality of production. The translocal can be understood as the intersection between articulation of the local (current issues, debates, performances) and wider informing contexts (cultural, geographical, political). Community radio broadcasts translocally at that 'moment when the local is stretched beyond its borders, whilst still remaining situated in the local', as Nico Carpentier argues: 'It is the moment where the local merges with a part of its outside context, without transforming itself into this context. It is the moment where the local simultaneously incorporates its context and transgresses into it' (Carpentier 2008: 6). A translocal articulation thus enables localised performance of community-specific debates and phenomena, which then can transcend the immediate context to attain purchase in wider contexts (cultural, geographical, political) while retaining its local integrity.

Somewhat unusually for US community radio stations, which often struggle for recognition and listeners beyond their regular supporters (often comprising a base of family and friends of volunteer practitioners), WRBH and WWOZ in particular are generally known and listened to in New Orleans, as suggested in my casual conversations while staying in the city. With its successful online streaming and branded website, WWOZ transmits curated New Orleans music alongside everyday moments of culture embedded in music shows but also via information about local gigs and neighbourhood events read aloud on air by a range of localised voices. In its transmission of similar moments of local culture found in articulations and discussions of local food, arts and theatre events and interviews with famous New Orleanians, WRBH mines New Orleans culture for a potentially wider audience while continuing to provide space for representation of under-represented ethnic communities alongside its substantial reading service for blind and other listeners articulations. Finally, in its determined focus on both the everyday of living with HIV and negotiating ongoing political and social problems of New Orleans's infrastructure, WHIV brings intensely local struggles to a wider audience, mobilising articulations of these struggles in ways useful to a larger politicised listenership.

NOTES

1. Station manager Natalia Gonzalez recognises that the word 'handicapped' in the station name is today considered offensive; she notes however that the broadcast call sign can be difficult to change officially. (From interview with Gonzalez, 13 May 2015.)
2. WRBH's 'The Newspaper on the Air' show broadcasts live reading of news from The Times-Picayune, Nola.com, and The New Orleans Advocate each morning between 7-9am, Monday through Friday. See http://www.wrbh.org/show/the-newspaper-of-the-air/, accessed 18 December 2018.
3. Interview with Gonzalez.

4. Interview with Gonzalez.
5. Interview with John-Hoa Nguyen, 20 May 2015
6. Interview with Joseph H. Louis Jeune, 13 May 2015.
7. Interview with Jeune.
8. Interview with Nguyen.
9. Interview with Jeune.
10. Interview with Nguyen.
11. Description from the WWOZ website, http://www.wwoz.org/about (accessed 10 September 2015).
12. From http://www.wwoz.org/about (accessed 10 June 2015).
13. Interview with Jorge Fuentes, 22 May 2015.
14. Interview with Fuentes.
15. Interview with Fuentes.
16. Interview with Fuentes.
17. Interview with Dave Ankers, 28 May 2015.
18. Interview with Ankers.
19. Interview with Fuentes.
20. Interview with Ankers.
21. Interview with Curtis 'Action' Jackson, 28 May 2015.
22. Interview with Jackson.
23. Interview with Jackson.
24. Interview with Fuentes.
25. From https://whivfm.org/about (accessed 7 August 2018).
26. See https://whivfm.org/schedule (accessed 7 August 2018).
27. From http://www.whivfm.org/ (accessed 11 September 2015).
28. Interview with Andrew Ward, 28 May 2015.
29. Interview with Dr Mark-Alain Dery, 29 May 2015.
30. From https://whivfm.org/about (accessed 7 August 2018).

Chapter Two

Transcultural Radio in Miami

Miami occupies the edge of the United States geographically and therefore politically, as a border city and destination for Caribbean and South American migrants, and as home to an established Cuban émigré enclave. The city's established and emergent diversity incorporates substantial Cuban and Haitian communities alongside transient snowbirds from Canada, New York and the US Midwest. Spanish is the first language in many parts of the city due to Miami's substantial and powerful Cuban community. The everyday normativity of Spanish is reflected in Miami's comprehensive Spanish-language media, which has proliferated in print since the 1970s, with multiple Spanish language radio stations emerging from the 1980s (Girard and Grenier 2008). Spanish-language radio today has an established presence in Miami; at time of writing there are three mainstream commercial Spanish-language radio stations broadcasting to the city, affiliates of the Spanish Broadcasting System (SBS) broadcast group, which owns Spanish-language stations across the continental United States and in Puerto Rico.[1] The city is also served by NPR station WLRN, which primarily broadcasts in English. Miami's other linguistic and cultural communities are represented in more localised and diffuse programming, and broadcast provision for Miami's established Haitian community is primarily provided by a small commercial station, WSRF AM. Miami's alternative arts and culture scene, geographically centred around the Wynwood neighbourhood in North Miami, has generated several online-only stations which foreground local artists and musicians, as does the low-power (LP) FM station Shake FM, which has a music remit.[2] In terms of community licensing possibilities, then, Miami's diverse linguistic and cultural context, served by a Spanish-language mainstream media sphere and a predominantly commercial broadcasting environment, sits in contrast with the New Orleans

Figure 2.1. Mural in Little Havana, Miami, April 2016. Photo supplied by the author.

context incorporating three community stations, all licensed for FM broadcast. Conversely, Miami's community programming comprises sponsored community shows broadcast on small commercial stations, and a dedicated schedule of Haitian programming at Haitian American owned station WSRF.

ACCOUNTABILITY AND ALLIANCES IN LOCALISED COMMUNITY PROGRAMMING

I identified two community-produced and presented shows broadcasting from two small local Miami stations during my research in 2016: *On the Frontline* and *The Colonel Matias Farias Show*. These, as well as Haitian American programme *The Morning Drive*, engage at a community-led register with their target groups, broadcasting live, debating current local issues and regularly featuring guest contributors to speak to these. *The Morning Drive* addresses Miami's Haitian American communities through linguistic delivery articulation strategies, which will be explored further in the second half of this chapter. *The Colonel Matias Farias Show* combines English and Spanish in discussion of local issues, while *On the Frontline* broadcasts mainly in English.

Miami's community-facing programming is not directly facilitated through community license provision, as with New Orleans's three dedicated and licensed community stations.[3] Instead, at time of writing, Shake FM was the only LP FM licensed station in the Miami area. As noted in the introduction, US cities offer very few opportunities for the establishment of new LP FM stations—the license category usually applied for by stations foregrounding a community remit—due to the higher demand on finite spectrum provision in US urban areas. Thus, in a competitive urban market such as Miami, 'community interests often take a backseat to commercial interests' which then 'maintain a virtual stranglehold on the radio market in urban areas' (Echert 2015: 238, 239).[4] Consequently, community-produced programming in Miami tends to be paid for by the community, which either pays a stipend or brings in regular advertising for the host station. This requirement means in turn that those community-facing shows broadcasting on these terms are produced by groups who have access to such financial support, whether directly (in a stipend) or indirectly (through reliable advertising). In April 2016, *On the Frontline* and *The Colonel Matias Farias Show* broadcast in evening slots from a compact one-story building in Little Havana which housed studio space for three small radio stations (880 AM, 1080 AM, 1360 AM). While these local stations hold commercial rather than community licenses, the show formats and production practices—dedicated to in-depth discussion of community issues and regularly including key community representatives—are characteristic of community radio.

In my conversations with Miami radio practitioners working on these community-facing shows, who were mainly based in local and public service sectors in their day jobs, I found that their shows regularly referenced alliances across culturally and linguistically distinct communities. These alliances are reinforced—at least in part—through in-studio discussions taking place on programmes such as *On the Frontline* and *The Colonel Matias Farias Show*, known by most as simply 'the Colonel's show'. *The Colonel Matias Farias Show* broadcasts at 7 p.m. Thursdays on 1080 AM and *On the Frontline* broadcasts right afterward at 8 p.m. on 880 AM from the second studio in the same building.[5] On a Thursday evening in April 2016, I joined 'the Colonel's show' in studio as tiny cups of excellent Cuban coffee were passed around and the hosts argued amicably on air with each other and invited guests, occasionally silenced—if temporarily—by the Colonel's authoritative tones. In the studio next door, a programme primarily dedicated to Miami's African American communities, *On the Frontline,* featured interviews with two local developers about their new building project in the neighbourhood and the ways they anticipated this would contribute to the community.[6] Producer-presenter Gary Johnson interviewed the developers with an affable tone, yet his questions were pointed and focused on the developers' accountability and responsibilities to the community.[7] Both *On the Frontline* and *The Colonel Matias Farias Show* regularly include several community representatives and other community members during their live broadcast, in addition to the host and the invited interviewee, so that the presentation style is communal—although ultimately the flow is directed by the show's host. Gary Johnson explains that *On the Frontline*'s remit is community empowerment through education and information provision:

> Our most important goal is to inform, and to educate. What we've experienced in radio and through social media is that the community lack education on issues as it relates to and affects their everyday life. So our broadcast really represents an opportunity to do fact finding. To get as close as we can to real information that actually [can] bring upon a change in a community. . . . Our programme is mainly around growing and educating and empowering a community.[8]

Some cross-pollination between the two programmes was evident by my observation, as two contributors from 'the Colonel's show' later came to join the guests in studio for *On the Frontline*. *On the Frontline* and *The Coronel Matias Farias Show* regularly invite community representatives and local politicians on air to field questions about current developments in their constituent communities, calling them to account on air during live broadcast in both English and Spanish and thus providing alternative, community-facing fora for discussion of community-specific events. *On the Frontline*'s Gary Johnson explains how alliances between the Spanish-language and African American communities are regularly referenced and discussed on the show:

> We're a predominantly Hispanic community [in Miami], and you need that relationship to be a consensus to help empower your areas. That's why we ally ourselves with some of our Latino brothers and sisters to discuss issues that affect us and that affect them as well.
>
> And a lot of times you'll find that we have a lot more in common, because you find people that really have some of the same concerns that you have and they speak up about it . . . Ultimately, we try to find a common goal so that we can still work together to help our communities.[9]

While at time of writing Miami has no radio stations broadcasting with a LP FM license with a remit for representing any of the city's marginalised communities, these two community-facing programmes are doing grassroots communication and outreach work under the infrastructural, labour and material conditions of community radio, produced and presented by volunteer practitioners in somewhat under-resourced studios. As Bruce Girard (1992) argues in *A Passion for Radio*, what can be determined as 'community radio'

Figure 2.2. Shopfront mural in Little Haiti, Miami, April 2016. Photo supplied by the author.

can be defined by the participatory nature of how this content enables ongoing interaction with the given community, or communities.

ARTICULATING HAITIAN AMERICAN SITUATEDNESS ON *THE MORNING DRIVE*

Over at WSRF AM in North Miami Beach, the daily morning programme *The Morning Drive* broadcasts banter, commentary and interviews in a vibrant and flowing combination of English and Haitian Creole; or as one regular contributor to the show calls it, 'Cringlish'. Haitian Creole is a Francophone dialect spoken in Haiti alongside French, Haiti's official language of record, and is also the primary dialect for many members of Miami's Haitian American community. WSRF broadcasts programmes for and by Miami's sizeable Haitian community in the surrounding neighbourhoods and in nearby Little Haiti, as well as online, and bills itself as 'the first and only Haitian-owned station in the nation.'[10] At the time of my Miami research in 2016, *The Morning Drive* was produced, researched and presented by Rhonel Cinous, Fabiola Charles and Rebecca Laratte, all in their twenties. Rebecca Laratte worked part-time at WSRF in other capacities while she coproduced and presented *The Morning Drive*. Laratte and Cinous were also studying journalism while coproducing and presenting the show.

The Morning Drive's mandate is to broadcast to the Haitian communities served by WSRF, as stated on the programme's webpage: 'Through this station we want to present a show that will bridge the young and older Haitian communities together. ANN ALE!!!!!'[11] 'Ann ale' means 'let's go' in Haitian Creole and can be heard as an energetic call to listeners at the start and end of each show, with Haitian dance music playing underneath adding to a celebratory on-air atmosphere. The show is more formally introduced by a promo each morning, announcing its 'new and different' approach to radio, prefaced with a French greeting then read out in English by the three producer-presenters in turn:

> Bonjour, you're listening to *The Morning Drive* on WSRF, 1580 AM. The *Morning Drive* is a young and new show unlike anything you've ever heard before. This show is for the older generation, the new generation and the ones in between. We are bridging the gap. Join us Monday through Friday. Stay connected: your *Morning Drive* show is on social media. On Facebook, Morning Drive1580. On Twitter, tmd1580, and on Instagram, themorningdrive. Ann ale!!
>
> (from *The Morning Drive*, 18 May 2016)

The show's flow between English and Creole terms and expressions is inherent to the programme's fluid and lively presentation, and hosts and guests alike move easily between the two languages. Regular guests are familiar with the show's bilingualism, with one contributor—festival promoter Mia Lopez—casually asking at the start of her on-air interview, 'you guys speak in both, right?'[12] The three young programme hosts say this combination of and movement between English and Creole is a deliberate and carefully considered choice. They feel that speaking in both languages enables them to engage with older and younger generations of the Miami Haitian community simultaneously, through a mode of expression everyone can understand and relate to. However, the three producer-presenters have been criticised by some conservative listeners for not speaking entirely in either English or French—the 'public' language of education and radio in Haiti. Fabiola Charles contextualises this reaction in relation to community expectations:

> When they [Haitians] have a programme, it's either French, all French, or it's in Creole, or they speak Creole only. . . . If it's in French, it has to be in French. And if it's in Creole, okay, it's in Creole. There's not really a mixture of them speaking both English and French and Creole. I think we're one of a kind where we speak all these languages.[13]

The decision to broadcast in a combination of the two languages provided a way for the three producer-presenters to communicate across the diverse groups in their larger Haitian American community, enabling them to be understood across generational divides. Broadcasting in both English and Haitian Creole also reflects their own experiences as members of their community: Rhonel and Rebecca grew up in Little Haiti, while Fabiola was born in Haiti and moved to Miami as a child. Through their unique blending of the two languages even within a sentence, the three are able to assert their own stamp on to the programme. Rhonel Cinous describes their decision as a way of claiming autonomy while reaching out to their peers:

> For us, I guess culturally, the youth or the younger generation of the Haitian community are marginalized . . . so for us it was kind of a defiant move: 'You've no choice but to listen to us now'. Not only that but we find ourselves in a position where we can help our peers, our elders, because there are a lot of things that they literally just don't know about. So we get to be defiant but at the same time show them we're doing something productive.[14]

This focus on younger listeners is also reinforced through regular updates on Facebook, Instagram and Twitter, in recognition that these are the primary platforms for younger audiences. Rebecca Laratte adds that 'when they heard our programme it was kind of weird because they never heard anything like that before, so it made the whole concept even more inviting, because it was so different.'[15]

Carline Faustin is coproducer of an educational programme, *Radyo Lekol*, directed at Miami's Haitian community, which broadcasts on both Miami's NPR station WLRN and WSRF, and also regularly contributes to *The Morning Drive*.[16] While Faustin appreciates the trio's ingenuity, she too finds the on-air combination of languages controversial, and explains how this fluid mixing of the languages, incorporating slang from both Creole and English, is defended by the three producer-presenters as designed primarily for the benefit of their peer group:

> There's three Haitian youngsters who are doing a programme where they speak English and Creole. And they get a lot of pullback from people like me and from other older listeners, that say 'that is not right! You either speak one language and finish the sentence or speak the other!'
>
> . . . As a teacher [*laughs*] it's like: this is not the proper way to use the language. I get to meet them Wednesdays when I go to the programme, and I say 'did you hear that feedback that woman gave you?' And they say, 'you don't understand. The programme is not for people like you.' It's for people like them. And then we'll laugh about it.[17]

In response to the trio's arguments in favour of broadcasting in 'Cringlish', and as a teacher with in-depth knowledge of Miami's Haitian community, Faustin sees possible repercussions for younger listeners, as she explained these to *The Morning Drive* team:

> I say to them, 'you know what? I can deal with you being able to do this, because I know you know the difference. But what about the other youngsters, who are just barely making it in school, and they hear you speak like that; they think it's the norm. And they will do that, and they can't get a job doing that.' And I'll create scenarios with them to start a conversation, and we have fun doing it.[18]

In addition to drawing in their peers, the ongoing fluidity in language in the show means they can reach more listeners overall, as Fabiola Charles points out:

> We speak Creole sometimes, to them [the older generation] so they can understand us talk, so they can relate, so they can give their opinions, and we go back to English so the younger generation can relate. So we're going back and forth, so the first community we have is the younger generation and the second community is the older generation. And we do have to merge these, so we do have to speak so that both communities can understand.[19]

Broadcasting through Cringlish thus serves as a statement of their practitioners' own Haitian American identity and experience while simultaneously functioning as a nuanced and transcultural means of communication. The use of Cringlish here can be considered as a strategy of what I have termed *accented radio* (Moylan 2013, 2018), which can enable articulations of practitioner standpoints and situatedness. My concept of 'accented radio' builds on Hamid Naficy's (2001) exploration of 'accented' cultural production, to describe a particularly transcultural mode of radio production. Transcultural radio programmes can serve as an alternative to often reductive top-down radio representations of marginalised communities. My use of this concept expands on Naficy's theory of accented film production as a migrant mode of transcultural production which signifies not only the 'accent' of those represented but also experiences of migration and movement which are then embedded in the text produced. *Accented radio* enables grassroots expression of community debates and issues and facilitates exploration of community identities and experiences in distinct ways depending on local and regional contextual factors. Yet accented radio in its fluidity also enables the producers/presenters and listener alike to transcend such binary distinctions between 'the local' and 'the global' or 'dominant/

mainstream radio' and 'alternative' radio. Instead, in significant part because of the broadcast nature and uses of radio, broadcast radio practices and content reflect, mine and reinforce the messiness and inconsistencies of the community everyday. What is broadcast is a show in process, rather than a finished product. Thus, accented radio can simultaneously represent and reinforce community identity from a community perspective, negotiate the effects of limiting and damaging top-down community representations in dominant media, bridge incomplete understandings between the home and host cultures and countries, provide useful community information and challenge ongoing internal community preconceptions.

An 'accented' mode of production takes on new and different significance when considered in radio forms. Along with Naficy, I deploy a critical mobilisation of 'accent' to consider and denote the ways in which the situatedness of the producer informs and infuses the cultural text produced. I argue in turn that in radio the accent is arguably particularly powerful as it is the only aesthetic and sensory mode of meaning production. Certainly, all of us speak with an accent; as Naficy points out, it is impossible to speak *without* an accent. Yet the presenters' accents most often heard on mainstream US radio—both public service and commercial—tend not to be immediately recognisable as belonging to a particular region. Certainly, too, the particularities of regional accents can be flattened out through education, travel or exposure to other experiences conferring cultural capital (as identified by Pierre Bourdieu 2010). Additionally, as paid work in radio (as with most media) is desirable, so too is it competitive—meaning that graduates from prestigious universities are more likely to be employed as mainstream radio presenters (see Mark Banks 2017 on middle-class graduates working in the cultural industries). The professionalised but less textured accent which results from the combination of such cultural capital with broadcast training is both accessible (because more widely intelligible) and authoritative, infused with the articulated confidence of a normative delivery style produced by a mainstream broadcasting institution. In their authority, such voices are universalising—serving to mediate the information relayed (news, arts coverage, music knowledge, etc.) through a curation or reportage microphone and in this process reframe what is being imparted in the dominant voice of a given society. In contrast, as I argue elsewhere in this book as well as previously (Moylan 2009, 2013, 2018), community radio programming with its greater capacities for marginalised community self-representation, enabled by an inclusive remit, produces a greater number of shows with regional and linguistically specific accents. Broadcasting from the Haitian-owned WSRF, *The Morning Drive*'s community-facing remit can be heard from the rallying call of 'ann ale!' at the show's very start. The immediate value of this lies in community-produced broadcast content spoken through accents and vernacular—and across two local languages—familiar to the target listening audience: Miami's Haitian American community. While controversial, the particular accented delivery and fluidity between English and Haitian Creole produces a cultural, linguistic and political bridge across generations of the Haitian American community. Community members may protest at the informal form of mixed language delivery but no one is left out.

For *The Morning Drive* producers, their own experience directly shapes and inflects their deliberate choice to move between the languages in this way, as Rhonel Cinous explains: 'We know how to speak English, we grew up in the system, we learned Creole from our parents, we learned French, but our experience . . . is the flow of it. That's how we speak daily'.[20] This fluidity not only enables the show to reach diverse members of the Haitian community with their varying expectations and experiences of radio but is an articulation of the practitioners' own situatedness: their positionality as Haitian Americans. As Fabiola Charles notes, 'With your parents you speak Creole. With your friends you speak English. And then sometimes you need a balance. Like sometimes you take the English word, you put it into Creole, and it comes so naturally now'.[21] Central to the political work of an accented radio text is how voice is deployed. In a radio text, voice narrates and narrativises, providing both structure and topic, and establishes a point of view and of identification for the listener. During a talk radio programme, when given sufficient time on air, voice can be deployed *in process* by a given speaker in the form of an individual narrative (Couldry 2010) which serves as a situated 'account, implicitly or explicitly, of the world within which they act' (2010: 7). The on-air community voice signals familiarity to listeners at emotional and political registers, which in turn reinforces inclusivity and community representation. The political function of the voice, heard and located in the particularities of the community accent, gains credibility when expressed and heard through the specific material context of community radio. The material conditions of many community radio stations (Lewis and Booth 1989; Moylan 2013, 2018) contribute to and shape a production context which facilitates inclusive practices. Due to limited provision of resources such as studio space and available recording equipment, programmes in community radio are often broadcast live rather than prerecorded. This 'liveness' produces a greater sense of immediacy and veracity, suggesting that what is being heard is unmediated, or at least less mediated, because what is being said is being said 'now'—at the moment of broadcast (Scannell 1991).

The Morning Drive epitomises the work of accented radio in overlapping ways. By dialectically articulating layers of Haitian American identity and experience through a deeply nuanced and locally particular fluidity of English and Haitian Creole, but within commercially resonant mainstream show formats, *The Morning Drive* negotiates its complex multilayered local listenership from the standpoint of its three producer-presenters. The layers of community referenced in these articulations reflect Charles Husband's (2005) recognition of the ongoing negotiations necessary in minority ethnic media to balance the diverse experiences and affiliations of community members where these diverge depending on generation.

I suggest *The Morning Drive* embodies the transcultural mechanisms of accented radio by facilitating these articulations of specific cultural, ethnic and linguistic subjectivities which are simultaneously shaped by the communication structures of the hosting locality (Moylan 2013, 2018; see also Vertovec 2008). In its fluid movement between Haitian Creole and English, *The Morning Drive* speaks both from and to the experiences of the first- and second-generation Haitian Americans who form the show's regular listenership, representing both these diverse experiences while reflecting the individual standpoint of each of the three practitioners. As Rhonel Cinous explains, 'Because we're first generation Haitian Americans, our experiences are going to be different from second generation Haitian Americans. But there are still people coming straight from Haiti who are schooling here. So I think feeding that new energy and the new

Figure 2.3. Photography exhibit, Little Haiti, Miami, April 2016. Photo supplied by the author.

groups into the show is very important'.[22] In this recognition, Cinous articulates not only the importance of speaking to a diverse range of groups within the community, even within a generation, but also his own experience with negotiating that diversity when presenting.

NEGOTIATING DUALITY IN FORM AND FORMAT

The ways in which accented cultural production negotiates between the local and the global are identified by Naficy (2001) as typifying a transcultural mode of production, in which the accented form produces meaning dialectically within dominant production styles even as the transculturally produced content comprises and enables a set of alternative localised perspectives. Thus, an accented radio show broadcast by and for a particular community can, and often does, draw on recognisable mainstream formats to situate content presented in the localised community dialect or language and foreground idiomatic talk and expressions. This accented quality and duality of perspective articulated comprises the *aesthetic integrity* of each show. In combining mainstream characteristics of commuter talk radio with their particular and idiosyncratic use of Cringlish—both when speaking among themselves in studio and when interviewing guests—*The Morning Drive* draws on both dominant and alternative radio forms. The show produces meaning dialectically, in order to appeal to young Miami Haitian Americans and first-generation Haitians at the same time, partly through the fluidity of language use, partly through topic choice and localised discussion. Alongside a deeply localised mode of delivery, *The Morning Drive* adheres to recognisable mainstream morning drivetime formats in its programming, with each day's show featuring a particular theme. The week begins with 'Motivational Monday' followed by 'Health Tuesday', which includes detailed health information specific to many in the Haitian American community, such as the importance of not indulging in too much rich and fattening traditional Haitian food favourites and of regularly consulting your local doctor. Wednesday's show demonstrates perhaps the most mainstream approach of the week with a 'Hump Day' theme; the colloquial name refers to getting over the midweek 'hump' but could also be read as a sly reference to the Wednesday show's focus on sex and relationship discussions and advice. 'Community Thursday' brings the focus back to community-specific issues and problems and Friday's show offers the most open format, which Rebecca Laratte describes as a 'chilled end to the week.' In addition to discussion of relationship issues, the Wednesday show also carries selective celebrity coverage that the trio feel is relevant to their peers. Tuesdays and Thursdays broadcast more community-specific content, including health care provision and financial advice, as well as a single-issue focus on a timely community problem. Additionally, *The Morning Drive* includes detailed traffic information and updates for commuters, central to the function of morning drivetime radio. Occasionally, an event-specific focus will supplant the daily themed shows; for example, just before the annual Haitian Compas music festival on 21 May 2016, the show featured advance festival coverage. On Haitian Flag Day, May 18, the practitioners urged listeners to hang Haitian flags in celebration of the day. The same show included coverage of former Haitian president Michel Martelly's visit to Miami, which was covered simultaneously live on the show's Facebook page and on Instagram, so that, as Cinous tells listeners, 'you were able to be there vicariously through us', adding that 'you know, we're all about bringing you guys access to places you may not be able to get.'[23] In providing and broadcasting coverage of this state visit, *The Morning Drive* team mediated the experience of Martelly's visit for interested members of the community.

Decisions around who produces and who presents are made organically, with the trio rotating research, event coverage and in-studio presenting among themselves. Each of the three can do everything if necessary; as Cinous says, 'I think each of us at some point has held the two hours by themselves'. He expands on the value of this:

> I think that's the most important thing we could have done to make sure we could all handle all the aspects: the sound board, handling the music, and then being the personality all at the same time. It is a lot harder than most people realize; I tip my hat to those people who do it regularly. We are our own engineers, we are our own DJs: we do it!
>
> I think being able to rely on each other and respecting each other is the biggest thing in our favour. . . . And it swaps. One day someone will say, OK, I'm taking the lead because I know this topic.[24]

Often, however, the show is presented by all three practitioners in studio. Together on air, Rhonel, Fabiola and Rebecca provide lively, relatable and often irreverent discussion and sometimes, debate. Often one will disagree with the other two on an aspect of the issue or topic under discussion, which leads to energetic banter as they argue the point through. In these debates, on-air disagreements can provide an additional point of reference or identification for listeners attuned to the issue at hand but holding diverse views. At times and where relevant, each of the producer-presenters draws from their own family experiences in their discussion of a given topic to enable and develop further understanding. Anecdotal talk about local neighbourhood food shops and restaurants also gets lively but never becomes explicitly critical, as the three hosts are very aware of their platform and the need to support local businesses. The show has a regular call-in component, regularly fielding callers with comments on the day's topic who are put on air in turn, usually briefly for each caller, providing a range of voices and perspectives.

Through this dual quality comprising a varied, potentially contradictory range of content expressed through Cringlish but framed through a commercially successful talk radio format and building on the programme's functionally descriptive title, which signals its uses of morning drivetime conventions, *The Morning Drive* embodies a negotiation between the local particularities of its (experientially and generationally diverse) Haitian American community listenership and a performed familiarity and ease with conventional and chatty morning drivetime radio. In doing so, it fulfils a central function of accented transcultural production, in which the

> dominant mode of cultural production exists side by side with the alternative and emergent modes, which serve other functions and can mobilise other values. Although it may circumscribe, influence, and modify those modes, it does not obliterate either their particularity or their alterity, for it needs both in order to flourish and to remain dominant. (Naficy 2001: 43)

By modifying, or customising, dominant modes, both on air and across social media platforms, particularly on the especially heterogenous Facebook platform, *The Morning Drive* deploys normative mediating structures to produce a recognisable site but also a point of departure for deeply localised and community-facing content. In this way, the show represents an interstitial subjectivity through performing its contradictions and layers; to be interstitial, Naficy suggests, is 'to operate both within and astride the cracks in the system, benefitting from its contradictions, anomalies, and heterogeneity. It also means being located at the intersection of the local and the global, mediating between the two contrary categories' (Naficy 2001: 43). However I would argue that local and global affiliations are today perhaps more mutually intelligible (through increased familiarity through proliferation online) and less contradictory than in Naficy's formulation, with the potential to be harnessed and repurposed to benefit local promotional aims, for example; local and global standpoints are performed and experienced at different registers and accordingly resonate separately. Occupying such a standpoint of duality enables and expands the capacity to negotiate, interrogate and challenge normative representations depicting Haitian American experience reductively. It is this embedded hybridity which renders *The Morning Drive* variously accessible and appealing.

ACCENTED RADIO AND AND RADIO'S MATERIALITIES

In critically considering and analysing *The Morning Drive* as accented radio, I draw on its qualities as *performance* and as *process*. Each show's *performativity* comprises textual elements such as voice (incorporating accents, cadence, delivery, tone); liveness; music and segues; discussion and interviews; impromptu phone-in contributions, show structure and timings. Performativity is additionally inflected by external structural factors such as the scheduled time slot, material conditions of the broadcasting studio and platform dissemination. Each programme's *process* is traceable through identification and analysis of individual production practices and how these are shaped by practitioner standpoint. The work of producing accented radio is 'multifunctional and integrated'—as Naficy describes transcultural accented film production (Naficy 2001: 46). Multifunctional, because the producer is often also the researcher and the presenter, and sometimes drives the desk too—as we have seen with the practitioners of *The Morning Drive*. Integrated because the producer and/or station personnel oversee platform dissemination of show content and, for producers with more time and access to equipment in the more resourced community stations, archiving of broadcast programme content. The means of production are usually held in the hands of only two or three people. Accented radio is thus often characterised by liveness, localised in-studio conversational content and local representatives speaking to current community issues, rather than 'experts' from outside the community.

Many community radio talk programmes are characterised by an informal presentation style and a fluid running order (because less hampered by advertising), leading to extended phone-in opportunities and a greater flow of in-studio conversation, dialogue and debate—all broadcast live. These characteristics are created in significant part by practices created by working with limited material resources, producing in turn (albeit mostly inadvertently) an aesthetic specific to community radio. This aesthetic, typified by liveness and greater listener access via the phone-in component, results in an increased sense of intimacy, as there are fewer layers of mediation between on-air speakers and listeners. These qualities of intimacy, liveness and fluidity comprise key components of accented radio.

The accented radio text can function as both an alternative and a response to mainstream representations of the community represented. Considering the accented radio text simply as a text of otherness, while potentially reductive, does acknowledge its dialectical relationship to mainstream representations of marginalised subjectivity. Where communities are chronically under-represented, the accented radio text can both build on these limited, incomplete representations and challenge them—either directly, through calling out reductive representations; or indirectly, by simply providing an alternative set of representations. This assemblage of community-led articulations of identity in a dedicated communicative space enables a sustained forum for discussion of community experiences and problems which is locally, linguistically and culturally specific to the community(ies) represented.

The Morning Drive is produced in-house at the WSRF radio station, which because of its location in North Miami Beach, a neighbourhood adjacent to Little Haiti, is well situated to represent and engage with the Haitian community from relative geographical proximity. The station's other programmes include Haitian news coverage in *News from Haiti*; a dedicated sports show, *Madame Gougousse Haiti Soccer Cup*; dedicated Haitian music programmes and a show on Haitian film, *Cinema Vérité*, produced and presented by the station's founder and manager Rose-Philippe Coriolan. Additionally, WSRF regularly organises community outreach activities such as food drives, which are then promoted on *The Morning Drive* and other shows in the schedule. *The Morning Drive* gains (further) credibility in Miami's Haitian American community through being broadcast at WSRF. By sharing content on Facebook, the trio also reinforces the show as part of WSRF's schedule, which means followers of *The Morning Drive* can easily make connections on the show's Facebook page to other WSRF shows—and with their listeners. Disseminating content across social media platforms also makes *The Morning Drive* accessible to those listeners who eschew analogue radio and primarily access radio online: they can either swap between platforms if accessing the show from streaming apps such as TuneIn or the WSRF website, or avail of Facebook's 'live' function which visually captures the trio in studio along with guests during the broadcast. As with all media forms and platforms, how people listen to the radio has become increasingly diverse and individualised. By my observation in various regional contexts, often older listeners and those who listen to radio while driving have more entrenched listening habits and may rely more on analogue access (depending to some degree on the technology available in their cars). Younger listeners are perhaps more creative in selecting and accessing diverse radio content from different sources and tend to be less conscious of geographical proximity, given they are accessing content digitally—even local radio. The producers of *The Morning Drive* consider that their on-air platform confers substantial responsibility to their multilayered community listenership, as Fabiola Charles explains:

> If you're talking about community radio, you're talking about what is affecting that community. And if you're a part of that you have to speak about . . . what's going on in that community. . . . When we're saying community radio, you have to let the voice of that community be heard, and that will give your platform a

certain standard, because it's not just a talk show or you're playing music; it's a specific purpose and . . . even when you're not on air they can email you, they can text you, they can message you something that they've seen and then that will give them the opportunity to feel they're important as well and that they're included.[25]

The elements of platform range and credibility when combined with the show's deep particularity of accented delivery, recognisable radio formats and deep access to community events and necessary information position the show as relatable but also crucially credible—in turn making its producer-presenters accountable to their community, a responsibility they embrace. Show content is therefore often directed by community needs, as Rhonel Cinous explains,

We are genuinely invested in our community, our families, our friends; we grew up here and didn't move away, so what happens in the community regularly affects us. So historically Haitians turn to radio for news, or to find out about the delegates/candidates that are running for election . . . so it is literally our job to provide this information.

So, doing our best to make sure the interests of the community are being brought forward. Even the professionals we bring in are for that exact purpose.

We allow what the community needs to dictate what we're bringing to them. So on Thursdays we do food drives; or on Thursdays again we bring in a financial advisor talking about credit repair, or education, whatever it may be, but we gear it towards the interests of the community.[26]

Cinous echoes Charles's earlier observation of a particularly Haitian reliance on radio, which suggests that *The Morning Drive* resonates especially deeply as a community forum and information point for members of Miami's Haitian community. Additionally, as he adds, 'they see us every day.' Charles concurs: 'At the end of the day we do go back to the community, because it's where we live.' She explains how their local proximity means they can discuss what's happening in the community from a community perspective:

For example, I live in the North Miami Beach area, so if something is going on in the North Miami Beach area, I better talk about it, because at the end of the day that's where I go home. I'm experiencing the same thing. So talking about it on the radio; the community is experiencing this; the whole community . . . I'm part of the community; although I'm on the radio, I'm part of the community.[27]

It is their recognition and negotiation of this duality, of being part of the community while also representing it on the radio, that informs the work of accented radio—and of community radio. This accountability can get critical, even pointed, as Cinous explains:

> And it's funny because they hold you accountable. We see these people regularly. As much as you think you can get away from seeing them. . . . I was at a Publix [a local Miami supermarket chain], and someone heard me ordering, and . . . and then believe it or not, they'll remember a topic we've discussed a year or two ago. First of all, I don't remember what topic you're talking about—it was like, four hundred topics ago . . . They hold you accountable. So they want to know, well, why didn't you talk about this. Or, you talked about this, [but] I didn't like how you did it.
>
> And it's always respectful. But because we like the same things they like; we like the same foods, so we're going to these places, we're bound to run into them. And again it goes back to that sense of family, you're their 'kids', so they feel like they have the right to correct you and say what they want. And it's always coming from a great place—but again, that's why we have to be very careful what we're presenting.[28]

He adds that community feedback can be detailed as well: 'But it's that they will call you out on it. They'll be like, what about that time, why didn't you ask him *this* question. You know what? It is our job, so next time send us a text or email us and I'll make sure we follow up on you'.[29]

Such involved, invested local community responses are arguably only possible because local community members do hold the show—and its producers—accountable. A further layer of community feedback speaks to the three practitioners as community representatives. On the show featuring Haitian Flag Day coverage, on-air discussion about crowding at the former Haitian president's visit included some description by the trio of some jostling in the crowds. In response, one caller suggested: 'People are looking at you as an example [of good behaviour when people are pushing in crowds.] People tune in . . . all I'm saying is, you've got to protect your style because people are looking at you as an example'.[30]

As described by its three producer-presenters at the start of each show, *The Morning Drive* builds a bridge across generations of Miami's Haitian Americans. In their felt accountability to their community, the show's three practitioners produce a programme that is inherently community-facing—culturally, linguistically but also geographically. While simultaneously broadcasting and sharing content across online platforms, the show also resonates with local Haitian American communities in Little Haiti and North Miami Beach. In examining and considering the particularities of *The Morning Drive* as comprising community-facing performance and a set of transcultural practices which combine in this alternative mode of accented radio, I am situating the show's three practitioners' deeply felt commitment to community recognition, as fulfilled through an arguably subversive reworking of mainstream radio formats within the specificities of live radio itself. It is the show's *radioness* which makes possible the appeal, the accessibility and intimacy of *The Morning Drive*: its immediacy, relatability and generated credibility emerge in part from a very particular and deliberate deployment of language, situated in the everydayness of morning radio. Voice, as explained, serves multiple functions—emotional, political, social, recalling familiarity and promoting inclusivity—and is perhaps harnessed by *The Morning Drive* even more profoundly, creating a fundamentally alternative form of community programming. The power of radio still lies in the voice and voices speaking—because live radio is an everyday medium, the value of voice remains transient, felt and experienced only in passing.

CONCLUSION: AN ACCENTED AESTHETICS FOR RADIO

The integrity of an accented radio text comes from its fidelity to its own representative project, as conceived, refined and reinforced by the producer and shaped by how she or he considers the show's relationship and responsibility to the given community(ies). The complexity of radio programme production (also incorporating presentation and research) derives from the multiple layers characteristic of the medium of live radio, as production contexts (institutional, material and technical, in relation to dissemination modes) combine with decisions around and expectations of the radio text produced (its structure and aesthetics). Each programme produced can be considered (and is usually experienced) in relation to its aesthetic integrity; meaning, these shows function as radio *qua* radio, hence this consideration of their *radioness*. My analysis of the cultural work of *The Morning Drive* additionally considers, then, the aesthetic integrity of this radio text. By *aesthetic integrity*, I mean the traceable ways in which a given programme consistently produces meaning and messages in ways integral to the inclusive mission statement of the programme. In framing this as 'aesthetic integrity' I draw on Banks's (2017) idea of and call for a sociological aesthetics 'that takes both subjective apprehension and objective qualities as being component parts of the value of cultural industry goods' as a means of doing justice to their integrity in producing meaning, and the cultural value of such meaning(s) (Banks 2017: 6).[31] What makes the radio programme text unique is that it has the capacity to embody characteristics of both the everyday and the deeply culturally specific—which of course can also constitute the everyday in particular and localised ways. Aesthetically, radio operates on a single sensory register: sound, which therefore becomes complex and multilayered for the listener. Everything heard initially from the start of a given show has equal significance, although an order and hierarchy is soon imposed. The show is introduced and held together by a presenter or perhaps two presenters, who serve as the authority and take the listener through the show's organising structure: the running order. A show's running order orients the listener and any dead air is immediately unsettling and disruptive.

In an aesthetics of sound, the voice is crucial and central. Perhaps especially in radio, 'there exists a unique relationship between voice interiority, and identity, and it is perhaps because of this that voice and speech are always associated with potency and magical power' (Naficy 2001: 121). In the voiced radio text, accent is what we hear—it is the shape of the voice. It carries conscious and unconscious connotations of region and language and also of education and socioeconomic status. The 'value' of each voice heard on air is ascribed by the listener to a position within a hierarchy based on socioeconomic subject positionality, in which dominant accents carry authority. Yet value is also carried by particular accents when they are known to us, which in turn generate emotional connotations of familiarity and nostalgia. Hearing a particular and specific accent which is nonetheless deeply familiar to us is reassuring—for radio, this can make the listener feel at home with a show. Where dominant universalising accents can communicate authority, particular accents confer both familiarity and consequently a deep credibility for those who recognise them. The community radio text thus performs a further and crucial function of community-led representation—at a political register which is emotionally situated by a familiar accent, which in turn confers credibility in the given show as a forum for voicing and debating shared experiences (Couldry 2010).

The depth of our attachment to and the credibility bestowed upon the voice lies in our belief in the power of its proximity and its suggested presence. Thus, as Naficy argues, 'sound is immersive, functioning to incorporate and unify the sender of the sound with its recipient.' Following from this 'there exists a unique relationship between voice, interiority, and identity, and it is perhaps because of this that voice and speech are always associated with potency and magical power' (Naficy 2001: 121; see also Ong 1982). For *The Morning Drive* listeners, the show retains the potent intimacy and power of live radio—whether accessed via WSRF's AM frequency or watching what's happening in studio on Facebook.

NOTES

1. See http://www.spanishbroadcasting.com/brands-businesses/radio-stations (accessed 1 August 2018) for Miami station details.

2. Phone conversation with station founder Peter Stebbins, 3 February 2016; see also 'New Miami FM Station Shakes Things Up', *Miami Herald*, https://www.miamiherald.com/news/local/community/miami-dade/article2171184.html (accessed 10 August 2018).

3. Miami's single low-power community radio license is held by Shake FM, broadcasting at 107.9 FM with a remit to broadcast alternative music. From the *Miami New Times,* 11 May 2016, http://www.miaminewtimes.com/music/miamis-six-best-local-radio-stations-8430288 (accessed 3 July 2018).

4. In contrast, it is considerably easier for community stations in rural areas to acquire a LP FM license, as there is less competition for the spectrum available (Echert 2015).

5. From January 2018, the 1080 AM frequency was taken over by Miami Christian station Radio Luz, who also occupied the shared radio station building: see http://radioluzmiami.com/ (accessed 3 July 2018).

6. My use here of the term 'African-American' is informed by Gary Johnson's description of his show's central community as being primarily 'African-American', as distinct from discussion of Haitian American communities in this chapter (Interview with Gary Johnson, 19 April 2016).

7. Details from my observation notes in both studios, 7 April 2016.

8. Interview with Gary Johnson, 19 April 2016.

9. Interview with Johnson.

10. This tagline appears on WSRF AM's homepage, https://wsrf.com/ (accessed 26 June 2018), and is a station ID message played at regular intervals during broadcast (as heard during *The Morning Drive* in April and May 2016).

11. From https://wsrf.com/shows/the-morning-drive/ (accessed 21 June 2018).

12. From *The Morning Drive*, 24 May 2016.

13. Fabiola Charles in group interview with Rhonel Cinous and Rebecca Laratte, 19 April 2016.

14. Rhonel Cinous, in interview with Charles and Laratte.

15. Rebecca Laratte, in interview with Charles and Cinous.

16. *Radyo Lekol* is a dedicated minute educational programme produced for two stations: for public service broadcasting (PSB) station WLRN Mondays through Fridays at 9.05 p.m. for twenty-five minutes and for WSRF on Wednesdays for a half-hour between 10:30 and 11a.m. (see http://www.wlrn.org/programs/radyo-lekol and https://wsrf.com/schedule-programs/page/2/, both accessed 1 August 2018).

The name *Radyo Lekol* translates literally as 'school radio' in a Creole variant of 'radio l'école'; école is French for 'school.'

17. Interview with Carline Faustin, 8 April 2016.

18. Interview with Faustin.

19. Charles, in group interview, 19 April 2016.

20. Cinous in group interview.

21. Charles in group interview.

22. Cinous in group interview.

23. From *The Morning Drive*, 18 May 2016.

24. Cinous in group interview.

25. Charles in group interview.
26. Cinous in group interview.
27. Charles in group interview.
28. Cinous in group interview.
29. Cinous in group interview.
30. From *The Morning Drive*, 18 May 2016.
31. With the recognition or caveat that aesthetic integrity and meaningfulness may not be valued by objective criteria—the one nit I pick with Banks's approach here.

Chapter Three

Social Justice Practices and Programming in Montreal Campus-Community Radio

Up to this point in this book I have focused my critical attention on the diverse ways community radio practices can articulate, perform and reproduce a sense of community, via inclusive strategies such as accented radio, to reinforce affiliation and belonging. I suggest that community radio enables and encourages a particular nexus of inclusive production practices, shaped by institutional and material factors, to produce content which resonates in locally, regionally and culturally specific ways. With this chapter's shift from a US to a Canadian community broadcasting context, I want to consider the production work of community radio from a different structural perspective; in particular, the ways in which production practices are informed and shaped dialectically by community station policy. Given that community radio's inclusive practices produce radio content that—as argued thus far—has its own expressive particularities, I seek here to connect individual production practices more explicitly with the context within which they operate, to consider in more depth how individual practitioners negotiate and balance the institutional context of their host station with their own felt responsibility for their show's community-facing remit.

A radio programme text is shaped by station policy guidelines and by the material conditions of production, both of which inflect individual production practices. Community station policies in turn contain, and represent, informing values of the station; in a circular relationship these in turn inform programme textual components such as spoken delivery, framing of programme items, interview questions and music selection, among other elements of a given show. In this way, institutional guidelines function as the scaffolding of programme production and station practices such as scheduling and the smooth everyday maintenance of a given station. A community station's remit and guidelines function discursively and politically as well as instrumentally, forming a liminal rhetoric between licensing policy requirements, possible and realistic station practices, and listener loyalty and expectations. Radio station policies—like the practices and content they refer to—are of course then additionally (and crucially) shaped by official and cultural discourses of their region and nation.

Canada provides for community broadcasting in part through allocation of campus-community radio licenses. These require that the campus-community station is housed on a university campus and must broadcast to, and provide opportunities for, interested students and local community members alike.[1] Campus-community stations avail of the university's material and infrastructural resources and must adhere to their host university's regulations alongside broadcasting policy guidelines for community provision. Montreal is home to two campus-community stations, both on FM frequencies: CISM, at University of Montreal, broadcasting primarily in French on 89.3 FM, and CKUT at McGill University, which broadcasts primarily in English on 90.3 FM. CKUT mobilises its campus-community remit in a consistently inclusive collective approach to managing its service and arranging the schedule, which comprises a diverse range of community-produced shows. Like Miami, Montreal is functionally bilingual in many parts of the city and many people speak both English and French on an everyday basis, although parts of the city were previously primarily Francophone, as in other parts of the French-speaking province Québec. Montreal has a history of both alternative media and radical community movements, a context which contributed to the establishment of multilingual community station Radio Centre-Ville in 1972. Members of Radio Centre-Ville describe a 'huge movement for social change' which developed in Québec in the 1970s: 'This movement gave birth to a multitude of organisations, ranging from community-based daycare centres, women's centres, youth centres, citizen's action committees to several groups fighting for the independence of Quebec (Foy et al./Radio Centre-Ville 1992: 50).

The development of community radio was part of this movement. Radio Centre-Ville is still broadcasting and the station's established working relationship with CKUT is discussed later in this chapter. It is worth noting that this social movement's emphasis on inclusion did not seem to incorporate questions of First Nations or Inuit community representation, but rather focused on Montreal's diverse—and disenfranchised—other communities. Building on inclusive practices and traditions from its inception, CKUT today maintains an equitable approach to programming and scheduling and provides ongoing and holistic support for its volunteer practitioners. An acute issue for community practitioners working on an unpaid voluntary basis is the need for ongoing station provision of supportive material and managerial

Figure 3.1. Outside CKUT radio studios at McGill University, Montreal, June 2017. Photo supplied by the author.

conditions to ensure they can continue to plan, research, produce and present their shows in their allocated time slot. Central to maintaining these conditions is station provision of necessary production space(s) and equipment alongside thoroughly discursive policy guidelines which enable practitioners to continue to work in a supportive and respectful environment. Each of these are key factors for practitioners which can directly inform their individual decision to continue producing shows for a given station.

COLLECTIVE PRACTICE: CKUT FM

In its guidelines and everyday practice, CKUT is broadly typical of community stations in its emphasis on balanced representation of multiple communities in the schedule and insistence on training and supportive everyday station practice in a context of insecure resources. Yet CKUT is also unique in its sustained commitment to collective station management. CKUT operates entirely on a collective basis in its management structure and decision-making processes and—unusually for any kind of radio station—does not have a singular station manager role. Decisions are therefore made collectively, meaning that the station maintains a bottom-up approach inherent to the ideals of the community radio project. While a collective approach is characteristic of the ethics of community-led work, this is difficult to sustain in practice, and all other community stations that I have come across in my research have individual station managers. Louise Burns, who oversees sales and administration at CKUT, explains how having a standalone station manager would contradict their inclusive policies and practices: 'We did have a station manager in the past . . . the thing is, if the principles that you're putting out over the airwaves in the content are democracy, inclusive, non-hierarchical; and you *have* a station manager, the complaint of the station manager was that they had all of the responsibility and none of the authority.' She expands on the value of collective decision-making, acknowledging that this process is dependent on the reliability of the people involved: 'I feel like decision making works when people really trust the people who are making that decision, you know; if you think this hiring committee is a really good reflection of the station, then you're confident that it's not politics or personal or favouritism or whatever'.[2]

Joni Sadler, CKUT's music programmer at time of writing, goes on to explain that there are positives and negatives to an embedded collective approach:

> It's really good in theory. It's often difficult to implement because it's dependent on personalities and people's level of—not necessarily dedication, but

people have different ideas about things and it's hard to get everything to—you know, be cohesive in a way that's going to move things forward. It takes a lot longer. If you're trying to plan something, or make decisions, because there's a lot more dialogue and group discussion, things take longer. And that's part of it.[3]

Burns describes how this can play out:

There's a lot of examination of process, a lot of revision, a lot of exercises or workshops, things to reinforce or to re-check-in to make sure everybody is on the same page of whatever it might be. So it's more process, and sometimes you have that tension [between] people who just want to get something done and the principles of democracy, lots of conversations about what the project is, is the project producing radio or is the project moving forward collectivity and democracy, and open in transparency. So that takes up a lot of energy and focus, and it's very much the preoccupation of the staff—but completely not the preoccupation of the programmers.[4]

Recognisable nuances of inclusive community radio practice and content are captured in the guidelines and practices outlined in CKUT's *Blue Book*, the station handbook incorporating general guidelines alongside the station's mission statement and overall remit. Guiding statements in the *Blue Book* broadly engage with three areas of attention: aesthetic integrity; community relationships and representation; and volunteer practices and rights. Each of these areas is substantially supported by statements reinforcing anti-discriminatory practice and content. In particular, CKUT's approach throughout the *Blue Book* consistently reinforces the station's commitment to inclusive practice across the areas of cultural and community representation, incorporating the complex areas of programme content and production work. As is standard practice with community radio policy, production guidelines emphasising inclusive practice necessarily shape production of inclusive content, but also address the politics of assembling and maintaining a broadcast schedule which facilitates balanced and equitable representation of the diverse communities served by CKUT's mandate. CKUT's mission statement is representative of community radio policy more widely, combining a commitment to community service through information, education and training. In addition, CKUT's *The Blue Book* incorporates a further commitment to providing alternative programming:

As a campus/community radio station, CKUT's mandate is to provide an essential service to those in the Montreal community whose needs are not met by mainstream community radio. CKUT functions not only as an alternative to the status quo, but also as a viable community resource. CKUT serves as a training ground for the community and student populations, and in doing so, provides

> an essential educational and information service to the greater Montreal community. (CKUT *Blue Book* 2016: 5)[5]

In incorporating and foregrounding provision for 'an alternative to the status quo' for Montreal listeners whose needs are 'not met by mainstream community radio', CKUT recognises the need for such alternative programming, which station management consider at both aesthetic and political registers. CKUT's programming policies draw simultaneously on principles of inclusivity and the alternativism of its mission statement to expand possibilities for creative scope in their scheduled programming. Stephen Barnard (2000) develops an understanding of 'alternativism' as inherent to community radio's independence from and otherness to mainstream forms of radio (an understanding reinforced by community radio's recognition as third sector media in EU broadcasting policy; Lewis 2008). 'Alternativism' is also a term harnessed to describe regular provision of 'alternative' music, particularly in US programming contexts where music previously perceived as alternative to mainstream music programming today forms its own fairly lucrative and successful programming stream.[6] In addition to such categorization, conceptually 'alternativism' can refer to alternative *modes of representation*, such as grassroots or bottom-up and/or collective community representation, alongside or as well as alternative *aesthetic content*, such as alternative music which itself forms multiple subgenres from alt rock to alt country among others. In its conceptual relativism and as a descriptor, alternativism can be usefully mobilised as an aesthetic criterion, building on an understanding of 'other than mainstream' to incorporate musical subgenres, artistic spoken work programming or art radio. In its policy and guidelines discourse, CKUT harnesses framings of alternativism in Barnard's sense as alternative to the mainstream, which in turn produces a discursive point of departure for programme content which encompasses recognition of the subjective, individual and affective qualities embodied in music and audio art produced for radio. Both the 'Music Policy' and 'Spoken Word Policy' guidelines enable space for creative expansion in separate ways; under 'Music Policy', the fourth guideline reads

> CKUT is mandated to explore all kinds of music. These can often include texts of artists exploring the most difficult facets of human experience and behaviour. . . . CKUT believes that artists often act as the emotional catharsis of our society and as such have a right to be heard. CKUT's programmers must be prepared to justify the airing of such material by identifying its redeeming factors and by providing a context. (*Blue Book*: 9)

This builds on guidelines which advise against playing discriminatory musical material including lyrics and including recognition of instances where such language is satirical; the above statement follows on from guidelines against offensive and discriminatory content and draws on these necessary caveats when encouraging creativity, in anticipation of possible clashes seemingly in service of creative production. The 'Spoken Word Policy' guidelines begin by combining recognition of the centrality of the aesthetic alongside the anti-discrimination rule:

> CKUT recognises and celebrates that the parameters of art are enigmatic and vary with individual perspective. However, CKUT will not in any way be a vehicle for the promotion of discrimination, hatred, stereotypes, abusive comments or offensive material. When material is broadcast that could be interpreted in such a fashion, programmers will acknowledge this interpretation on-air and will at all times contextualise the material through intelligent explication, analysis and responsible discussion. (*Blue Book*: 10)

This statement confers the expectation of responsible interpretation and mediation of creative material by practitioners, producing a station-supported framing of such material and encouraging individual practitioner agency in doing so. By foregrounding the aesthetic but with anti-discrimination measures immediately following, this statement particularly exemplifies the multiplicities of the community radio text.

Building on this dual understanding, the *Blue Book*'s 'Free Chat' guidelines recognise how representational strategies can function within delivery style and actively advise on ways to ensure that community perspectives are foregrounded and that sensitive issues are represented respectfully and with integrity. These open-ended suggestions sit alongside policy statements which focus primarily on representative production strategies. The 'Free Chat' guidelines foreground inclusive on-air discussion suggestions which emphasise reflective practice during in-studio conversation—such as, 'Am I speaking from first-hand experience, or opinion?'; 'Have I researched this topic and am trying to inform the listener, or is it just my opinion or an attempt to get my voice heard?'; 'Am I interrupting others on the mic?' and 'Am I speaking for others?' (*Blue Book*: 36). Such questions are effective in the in-studio moment for checking and correcting modes of delivery which unconsciously replicate a status quo celebrating an individual presenter's voice as dominant—both in terms of what is expressed (if the presenter speaks more than the guests, a single message is emphasised in a top-down delivery mode) and who feels enabled to contribute. Following the above question which asks if the presenter is speaking for others, the following is offered to exemplify what bad practice might sound like: 'A white western woman or man cannot speak for Muslim women in [T]urkey or women wearing niqab in Quebec'; this usefully connects delivery practice to the possibilities of reinforcing wider cultural assumptions. Subsequent sample questions are followed by explanations of how delivery style and habits can shape how meaning is (re)produced through these practices, to draw on one example:

> Am I monologuing, preaching or falling into a rant?
>
> Does it sound like I assume the listener needs to be told what to think, and am I fulfilling that role? Am I using the airwaves as my own personal soapbox or cultivating my cult of personality? Be wary of reports that are more focused on yourself than the subject being covered.
>
> Do I find myself needing to comment on everything that happens during the show?
>
> If you are doing any of these things you probably need to stop or find a new way of presenting information. (*Blue Book*: 36)

In addition to their use value for in-studio practice, reinforcing persistent connections between speaking directives and their wider societal and political instrumentality cements recognition of how these practices (therefore) shape meaning production. The 'Free Chat' section concludes with the following 'last words':

> The role of a host is to facilitate conversation, bring guests on the air, present information and keep the program on track. The role of a host is not to provide commentary and opinions on all topics presented—that's the role of guests that come on the show, from relevant backgrounds, organizations, etc. (*Blue Book*: 38)

This short set of guidelines can be said to function dialectically, here and in the excerpts above, by combining pragmatic practice guidelines with accompanying observations on how inclusive practices can be mobilized discursively by drawing on delivery styles as examples. CKUT's *Blue Book* is representative of community radio's institutional commitment to inclusivity (and is itself a product of collectively agreed-upon practices and terminology) but also of radio practices more broadly—of the mixture of consistency and reliability which good radio requires to ensure listener loyalty, alongside radio's capacity (because so trusted) to provide a discursive space for in-depth articulation and discussion of a wide range of experiences. In their *Blue Book*, CKUT incorporates this dual understanding of radio practice and ways in which this duality informs the creation of programme content. The station's practices and the ways in which these facilitate inclusive and equitable programming can be usefully considered through the hermeneutics of what Mark Banks (2017) terms *sociological aesthetics*: in which coherently inclusive production *practices* shape the aesthetic integrity of radio *content*, with policy attention to *form* (responsible talk radio practices) enabling the *function* of community-facing content. In regularly and actively reinforcing

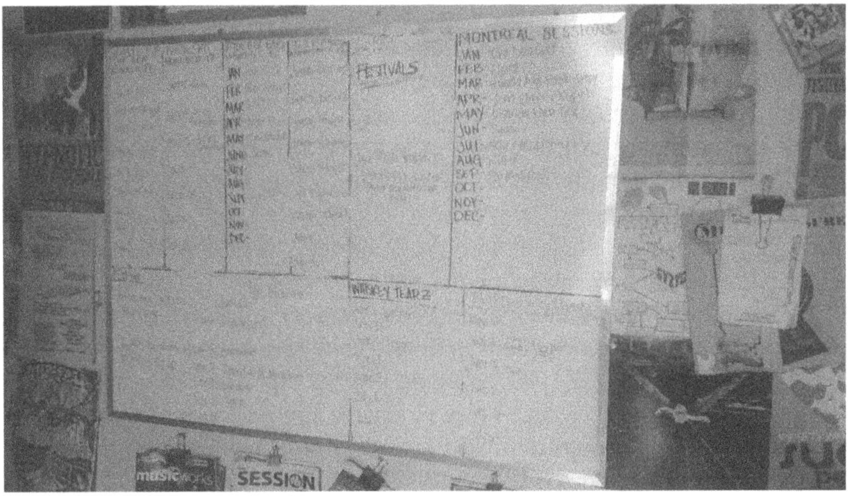

Figure 3.2. Station scheduling practice in process at CKUT, Montreal, June 2017. Photo supplied by the author.

an inclusive approach in their station practices and principles (as documented in *The Blue Book*), CKUT's approach to practitioner oversight and station management (overseeing producers but also determining and refining the schedule) can be read as comprising an ethics of community practice, while exemplifying what Banks describes as a

> notion of practice [that] therefore refers not to an unrecognised, individual ability or performance but to collective recognition and regard for the standards of worth inherent to the activity in question. What unites people in a practice is a shared commitment to the practice, which is recognised as unique and distinctive, with a set of internal standards of excellence that one must seek to uphold for the good of the practice and for the good of practitioners themselves. (2017: 46)

Such a collective practice's value then in turn resides in its value to its community: 'a practice always relates to the wider recognition of how individual skills and abilities are directed towards the "higher" good of the community' (Banks 2017: 46)—a principle central to community radio's inclusive project.

EXPANDING DIVERSITY: SUPPORTING THE WORK OF COMMUNITY REPRESENTATION

'Community access' programming is described in the CKUT *Blue Book* as providing for 'those not adequately served by the mainstream media'; serving as a '"voice for the voiceless" in addressing issues of concern for specific communities'; promoting for the wider Montreal community 'an understanding of these under-represented communities and the issues they may face' and representing 'a wide range of different or differing views from within those communities' (*Blue Book*: 7). CKUT's stated commitment to community-facing programming is further evidenced in the station's broadcast schedule and the wide and eclectic range of community programmes it incorporates. CKUT differentiates between station-produced community-facing programming and those shows made entirely by community practitioners themselves, which comprise the majority of the station's broadcast content. 'Spoken word' programming is defined as focusing 'on specific issues or topics' and representing 'news and perspectives from community groups in the Montreal area' (*Blue Book*: 9). CKUT's schedule also incorporates arts- and music-based shows with a Montreal focus alongside mixed community-produced programming from groups defined variously through shared cultural, experiential, linguistic and/or political affiliations. In spring 2018, CKUT's extensive range of community shows included *Amandla*, an African news programme produced for Montreal listeners; *Desi Beats* representing Montreal's South Asian communities; *Dykes on Mikes*, representing Montreal's lesbian communities; *In the Motherhood,* focusing on parenting from a feminist perspective; *Montreal Arab*, featuring arts and cultural content for Montreal's Arabic communities; *Native Solidarity News* and *Nipivut*, representing Montreal's First Nations, Métis and Inuit communities; *No One Is Illegal Radio*, representing groups actively engaging in resistance, including migrants, refugees and First Nations, Métis and Inuit peoples, among others; *OWL* or 'Older Women Live'; *Prison Radio*, focusing on content 'directly collaborative with people who are currently incarcerated'; *Queercorps*, focusing on a 'feminism from women of colour perspective'; *Treyf*, representing a Jewish perspective; *Under the Olive Tree,* focusing on issues around the Palestinian occupation; *Venus*, representing 'feminist, queer, female, trans, and gender non-conforming gender identities' and *VOKO* representing the city's Korean communities.[7] Provision for McGill students and the university community includes *LegalEase*, which focuses on legal issues and is produced by McGill law students and *All Things McGill* directed at the wider university community. CKUT's wide-ranging and eclectic range of music programmes foreground culturally specific and locally produced music across multiple genres, while the CKUT-produced show *Radio Is Dead* produces alternative (as conceptualised above) radio content ranging from spoken word to soundscapes and encourages collective creative participation.

CKUT's particularly diverse range of programming is representative of how the idea or concept of 'community' is conceptualised by the station in multiple inclusive ways. As an example, in response to my question about how the station enables self-representation of 'marginalised' communities, Burns asserts, 'That word: marginal, marginalised; that's definitely where we hang out. And if it's a metal show, a hip-hop show, it's being presented by those people who are, like, right in there, upfront and centre'.[8] This reinforces CKUT's commitment to ensuring community representation is on the terms of the given community itself: by this logic, a specialist music or culture show is best produced by people already living and working in that subculture, as Burns suggests. CKUT sees value in unique programming when considering the possibility of adding a new show and when considering changes to the overall schedule, or 'grid', as Sadler explains: 'We try and think about what's represented on our grid already and what's represented in the bigger media landscape, and if it's filling a niche that's really, really unique, then to us that's still super important'.[9] CKUT's diversity of programming is further framed through the station's sense of its own alternativism, which inflects the station's inclusive and expansive capacity for recognising communities as determined by affiliation. For example, avant-garde radio programming is incorporated into the CKUT schedule as part of this commitment, alongside communities determined by cultural, experiential, musical and political affiliations.

The station's collective practice is also reinforced by the way programming decisions are made by practitioner-led committee rather than by a management committee; this is, as Sadler confirms, 'a pretty uncommon model as well . . . At the end of the day, we're not the ones who can unilaterally make a call. Like, does the Haitian community get this slot, or the theatre group get that time slot'.[10] Burns expands on the criteria for these decisions: 'And those decisions are made outside of financial concerns. . . . Our major asset, which is our programming, is completely in the hands of volunteers.[11]

CKUT additionally expands community access and opportunities to produce their own shows by incorporating rotating shows on several timeslots on the schedule grid. This practice expands the station's capacity for community representation by further extending the number of communities represented and simultaneously reflects their commitment to collective practice at the level of show production. Sadler describes the value of rotating shows for the station:

> It does a couple of things. It brings in new voices; it allows people to get on the air faster, they don't have to wait for a permanent time slot to open up; and it's also good for people who are really busy and can only do a show every once in a while, or they travel and they're here once a month. And so it allows a lot of flexibility.[12]

The slots for rotating shows are thematic; while each show is distinct they may have overlapping approaches for community foci. On Mondays at 7 p.m. *Dykes on Mykes* and *Lesbo-sons* broadcast on alternative weeks. On Wednesdays at 6 p.m. the show slot is shared by *Audio Smut*, *In the Motherhood*, *OWL* and *Spitfiyah* and on Thursdays at 2 p.m., the following shows are broadcast on alternating weeks: *Tranzister Radio*, *O Stories*, *Radio Moon* and *Art Scene & Heard*. On Fridays at 11 a.m. *Shtetl on the Shortwave*, *LegalEase* and *Prison Radio* broadcast on alternating weeks.[13] Alongside and sometimes within this scheduling practice, the rotated shows themselves are collectively produced, which further reduces the pressure on individual practitioners, and importantly, functions to destabilise a hierarchical structure within conventional radio production practices, as Burns goes on to explain: 'I always say, you try and make it like someone can be the host of the show the first day, the guest on the show the other day, the listener the third day; those lines are completely broken down'.[14] At a wider register of production labour, this practice of programme and practitioner rotation exemplifies the tendency in community radio for a given show's producer to simultaneously fulfil the role of researcher and on-air presenter—further reflecting the realities of volunteer radio practices.

CKUT's provision of a supportive environment for volunteer practitioners means that many of their community shows are long-running and well established. At time of writing, Ayesha Baig has produced and presented *Desi Beats* for over four years and Sawssan Kaddoura has presented and produced *Under the Olive Tree*, a show dedicated to coverage of and discussion about the Palestinian occupation, also for four years.[15] In interview, both Baig and Kaddoura repeatedly emphasise their recognition of and trust in CKUT's provision of a supportive environment for their respective shows.[16] Both Baig and Kaddoura relate the support they've experienced to CKUT's stated mandate, which they see reinforced by station management practices. Particularly important for both practitioners is CKUT's recognition of their respective autonomy as producers: this enables their independence as practitioners and demonstrates respect for their judgement in deciding on show content. Baig describes the multiple ways in which CKUT provides support: 'I could understand the mindset of CKUT right away. . . . I really like the way they stand by their ethics, I've attended their meetings and they're tough people and they're very hardworking'.[17] She relates her show's project back to CKUT's mandate: 'It's important to me that it [the show] does justice to CKUT also. CKUT has a lot of respect in the community. . . . We [volunteer practitioners] feel loyal to it'.[18] CKUT's established support for its volunteers is an important consideration in the political economy of the cultural labour of community radio production. Practitioners feel supported by the station, which in turn is a further incentive to keep broadcasting their individual shows—other elements of their lives permitting. Both Baig and Kaddoura describe feeling supported at the level of production; Kaddoura sees this across CKUT's practitioners: 'We have complete freedom: CKUT never told us, you need twenty minutes for politics, twenty minutes for culture—and I think that's something that applies to all the shows. People have complete control over their shows'.[19] Louise Burns attests to CKUT's practice of providing infrastructural support while enabling autonomous production: 'We just make sure the microphones and turntables are working, and sometimes get them accreditation for certain things, but largely they're completely running their own content, everything'.[20]

Given that community-facing shows undertake the work of community representation, a project which necessarily incorporates ongoing consideration of which groups within a community are being represented, practitioners must continually negotiate the questions of: How can representation of multiple community groups be balanced equitably? What responsibilities does the practitioner have to their given community? *Desi Beats* and *Under the Olive Tree* each fulfil a representative function; the communities each show represents are self-defining in different ways. Ayesha Baig sees the project of *Desi Beats* as comprehensively representing South Asian culture, in all its manifestations and as authentically as possible:

> I want to represent every South Asian person living in Montreal and in Canada in the most justified way. Our true culture, our true poetry, our true artistic side should be shown to the world. This is my integrity and nothing else. But that requires a lot of work. You have to dig in a lot. Because not everything that's very authentic is going to be shown in a Bollywood film.
>
> The main thing is justified representation of all the South Asian countries so that nobody feels left out. That's very important for me, you know? You tune into a radio show and you listen to your own language and no one is playing that song—that's a nice feeling![21]

Baig describes the multifaceted appeal of such curated community programming, where the appeal to community members lies in the specialist cultural content itself (arts and events coverage, interviews with community artists, community-sourced music) but also the fact it is being introduced by a fellow community member, with the value in hearing 'your own language' on air along with the song 'no one else' is playing. In *Desi Beats,* community representation is situated alongside cultural curation of an array of South Asian community content. Baig sources South Asian music and cultural forms (film, poetry) to broadcast on *Desi Beats* with a view to educating a wider Montreal listenership beyond the city's South Asian communities. As she points out, South Asian listeners are often already familiar with the cultural content she plays: 'a lot of my listeners—through phone calls, through Facebook page visits, I realise—are Francophone, they're Arabs, not just from my community. Because *we* know where to find things. It's the rest of the Montreal listeners who wouldn't know how to look for the information on what's happening'.[22]

Broadcasting this cultural content appeals to other South Asian listeners but simultaneously provides a rich vein of South Asian culture for those unfamiliar, while challenging circulating stereotypes in the process:

> A lot of my South Asian friends have always said whatever people think of us is just not true. We are more than just Chicken Tikka or Bollywood music. A lot of my friends have said that to me. But the problem is a lot of the people of Montreal, a lot of the listeners of CKUT wouldn't know that.
>
> As a presenter I thought it was my responsibility to do my homework and show a bit of the other artists we have; let's say a small artistic film from Bengalese cinema . . . why not. I might have to dig in but then it's such a good representation of the community. And then, people need to know![23]

Many of CKUT's community shows have dedicated and loyal audiences, as Louise Burns explains: 'The community shows tend to have audiences that are particular to them. They're personal in a different way . . . they're speaking to people whom they actually visualise [via on-air shout-outs], to the extent that they're always surprised about the people who are listening that they don't know personally.'[24]

Desi Beats is broadcast on Sunday evenings between 6 and 7 p.m. For Baig, the show's name had to appeal not only to Montreal's South Asian communities but to a wider Montreal listenership, so she considered the show's name carefully:

> It had to be something modern but also something from the community. So 'Desi', it's a word for the South Asians. That's what we use. So a South Asian person listens to that, they understand it right away. But another person would not. So they would connect with it right away.
>
> And 'beats'—it just went with the flow. So I had a couple of names, so I asked my South Asian friends, I asked my white friends, I asked my Mexican friends, French friends, what is the best name that you like. And the majority went with this one.[25]

For Baig, CKUT in particular is the ideal home for her show and its inclusive project, as she describes it:

> I've been offered paid radio shows, a couple of times from other places. But I just didn't feel that connection. And also, accessibility matters a lot. McGill

Figure 3.3. Inclusive signage in Canada's capital Ottawa, June 2017. Photo supplied by the author.

> matters a lot to me. And being a McGill radio station, accessibility was there to begin with.
>
> Did I know it was a big commitment? Yes. Was I happy to do it? Yes. I'm still happy. I just hope that it survives.[26]

Baig's comments demonstrate the substantial commitment (typical of community radio practitioners) involved in regularly producing a weekly radio show, particularly when the show goes out live each week. She strongly believes in the form of community radio—from both a listener's perspective and for its capacities for community representation—and considers this in her conceptualization of *Desi Beats*: 'My show, I feel, is owned by the community. I don't own it'.[27]

BROADCASTING SOCIAL JUSTICE: ACTIVIST RADIO

CKUT's maintenance of and insistence on a social justice mandate, traceable in the station's diverse structural practices and declared support for social justice issues, makes it a supportive environment for the production and broadcast of programmes foregrounding social justice areas. CKUT's insistence on collective work practice is itself characteristic of social justice movement aims in its decentring of hierarchical work structures and decision-making processes which reinforce social inequity. When the community represented defines itself as a community of political affiliations, questions of representation become more complex again. The Palestinian solidarity show *Under the Olive Tree* is described on CKUT's website as 'Canada's only radio show entirely dedicated to Palestine. The show covers various news and topics related to Palestine and the Palestine-solidarity movement in Quebec, Canada, and around the world'.[28] The show's producer-presenter Sawssan Kaddoura explains how the focus on Palestinian solidarity necessitates a politicised perspective to all their coverage: 'None of us were Palestinian so we call it Palestinian solidarity. Because the Palestinian cause is very specific, very political, politics goes into everything we cover'.[29]

Kaddoura feels CKUT's established commitment to social justice makes it an especially appropriate home for *Under the Olive Tree*:

> It falls under the mandate of CKUT. Because CKUT is very much a station that is linked to grassroots, social justice, so our coverage of Palestine is very much in the same vein.
>
> The people who are listening to the show are the people who are listening to CKUT in general. And if you know about CKUT, you know these are people who are looking for independent news. CKUT has a mandate for caring about social equality, and fighting oppression. It's very much part of CKUT identity anyway, so the show also falls right into that.[30]

It is evident here that CKUT's ongoing engagement with areas and issues related to social justice is central to the station's cultural currency. The fluidity of community radio programming can facilitate a greater depth and breadth of articulation of multiple political perspectives, potentially producing an array of political standpoints for the engaged listener to identify with. Affiliations around social justice issues can be produced and reinforced through a nexus of nodal points, each relating to a particular political struggle or movement. People active in social justice contexts may affiliate with multiple movements but often emphasise one in particular as a focus for their activism and actions. While *Under the Olive Tree* foregrounds the single issue of the Palestinian occupation, Kaddoura recognises the show's audience may come to listen from a variety of social justice affiliations. Writing more broadly on how cultural practices can be activated for social change, Banks argues (2017) that 'practitioners—through immersion in the materially specific tradition of the practice—become *agents of change* through the constant refinements, challenges and innovations that are introduced as the limits to the practice are explored' (Banks 2017: 46, italics mine). The institutional and material conditions of community radio which combine to facilitate and produce a greater capacity for articulations of community positionality also enable spaces where the complexity of these political nodal points can be debated and discussed in material ways particular to the strategies of the given show. Thus, through such materially situated strategies, programmes such as *Under the Olive Tree* can enable activism on radio, through raising awareness but also by facilitating debates. Through the established cultural and political currency conferred by its social justice mandate and practices, CKUT actively facilitates the activism of *Under the Olive Tree* and other shows on the schedule which also foreground on-air activist work. Further, *Under the Olive Tree* produces an activist broadcast space precisely through aesthetic and articulation strategies specific to talk radio, and again particularly to community talk radio. Christina Dunbar-Hester (2014), writing about US low-power activist radio, suggests that radio activism in the current moment can be understood as 'issuing from embedded practices of community media production and pirate radio' further contextualised by overlapping media regulatory frameworks and societal circumstances enabling (legitimately or otherwise) its ongoing production. At the same time, 'radio activism was also understood to promote community-level self-determination and resist corporate encroachment' (Dunbar-Hester 2014: 28). That is, activist radio must be considered both as *activism* and as *radio*; so activist radio produced for broadcast on CKUT can be understood as simultaneously framed and informed by the station's own social justice remit which certainly facilitates such community-led self-determination.

Under the Olive Tree is broadcast between 11 a.m. and 12 noon on Thursdays, a schedule slot which lends itself to talk radio and in-depth discussion. For activist radio, the talk radio form provides a necessary credibility which in turn, as a recognisable format, produces cultural and sometimes linguistic intelligibility. Within an overarching project of raising awareness and providing information, Kaddoura notes that she specifically incorporates coverage of Canadian and Montreal contexts into her production practice, in recognition of the need to present political information to a diverse listenership, 'Because we are not in Palestine, we are in Canada, so you have to take the context in what you are covering'.[31] A further focus of *Under the Olive Tree* is political education, as she explains: 'There's a big factor of education, of raising awareness about Palestine. We are in a country where they [the general public] don't know much so they need to be given the information. So we assume no one knows. There's this Canadian audience, this solidarity audience, that we want to reach out to'.[32]

For Kaddoura and *Under the Olive Tree*, the political need for linguistic intelligibility is in tension with the show's project of specific information provision and alongside raising awareness:

> Language can be a barrier because we talk in English but we are presenting to a community that doesn't speak English, and we want their voices to be heard.

> I want to have more Arabic but our goal is very much to reach out, that's the thing: you're not only speaking to that community but a really big part of our mandate is to reach out to people outside of our community to raise awareness about what's happening with them. So then we are more limited to having English or French because I need to reach out. . . .
>
> But there is a value of also listening to the Arabic language, especially in the case of that language, because again . . . you only hear it when people are screaming or something [laughs]. It's good to actually hear it in a normal discussion context, how people actually talk.[33]

Alongside the need to balance presentation of information, interviews and other coverage in a nuanced way relatable to the Palestinian solidarity community but also accessible to a wider, less well informed listenership, is the need to be representative in relation to language:

> When I interview Palestinians, I have to conduct the interview all in English. Most Palestinians [don't speak English], so there's a lot of voices that I can't bring to the station. . . .
>
> So you're still limited to people who speak English. And they're still very good, people involved in grassroots, whatever, but I wish I could include all these voices. Because there's a lot of people who are doing great things, but they don't have [the ability to] interview [in] English.
>
> I only did it one time, but there was a woman from Gaza . . . I interviewed her in Arabic. There are so many voices like this. There are so many things that I want to cover, but because of the language . . .
>
> It takes a bit more time because you have to interview in Arabic and then you have to translate it. So one, it takes double your time because you have to play the original and then translate it, and then you have to format the clip.[34]

Negotiating the broadcast of community first languages alongside English (as a working regional lingua franca) is a familiar challenge for community-led shows where the community(ies) represented use their first language in everyday contexts. However, the show also seeks to broadcast activist content for listeners affiliated with the Palestinian struggle, who may only understand English—requiring a further layer of language balancing on air. As Kaddoura adds, crucially: 'I want these testimonies to be heard and understood.'

Producing activist radio incorporates additional layers of labour, as is clear in Kaddoura's reflections on her production practices and the ways in which she conceives of *Under the Olive Tree*'s remit. The self-imposed mandate for community representation felt by many community show producers becomes more acute when the community represented identifies itself as politically affiliated with a cause which is perceived by some in the mainstream to be controversial, and certainly the Palestinian solidarity community is a politicised community. In this case, care needs to be taken to protect the affiliated community of listeners as well as show guests and contributors. The work of community representation therefore also becomes more acute, incorporating a particularly urgent need for accurate information provision, for community-building and simply to maintain a reliable presence on air. Kaddoura explains how these all overlap in practice:

> [O]ne thing also is very important for our show. With everything we mention, we make sure we mention we have resources [information/articles, etc.]. Usually we always mention resources, we really pay attention to not sit there and just rant.
>
> We always say the real point is education, so we always go back to the source and we always explain the facts. When you're on radio you have to be very responsible, you can't just say a thing and it turns out to be wrong. When it comes to a Palestinian show you have to be more vigilant. Because you might have people just waiting for you to say something wrong to discredit the entire show. Waiting for you to just say one thing wrong. So we are very careful.[35]

Reflecting on the responsibilities involved in producing *Under the Olive Tree*, Kaddoura discusses her work on the show in relation to the show as activist radio:

> As is probably the case for a lot of people who work on causes like this, I like the fact that I am working on a cause that is important. I don't know, it's the same reason why activists are activists, I just want to contribute to these issues that are very important. So I think that's part of it, if you do a show about Palestine it's contributing to this very big social justice movement that is happening, that is important. . . .
>
> It goes back to the concept of responsibility. Like there's no way that I would stop doing the show, because if I don't do the show there is no one to replace me and the show will go off air. And that's not an option. So it goes back to the idea about responsibility, because it's so essential right now to raise awareness about this cause, there's an urgency to it, more than just representing a community that exists in the city, there's an urgency to keep us on air to keep talking about this, until we solve the problem. [*laughs a little*]
>
> These shows, these community shows, it's the responsibility of us to keep it going. If we don't have anyone else to host it, the show will go off air. They [CKUT] will try to help, but we are responsible for keeping that slot, that show on air. It's never an option to risk having the show stop, because you need that space to stay clear.[36]

Here Kaddoura summarises not only her own substantial work researching, producing and presenting the show, but articulates a further concern: the importance and necessity of keeping *this* show—as an activist broadcast space dedicated to sharing information and providing context on the Palestinian occupation—live and on air.

As Kaddoura notes above, CKUT broadcasts several programmes which incorporate an activist approach—their inclusion contributes to CKUT's overarching commitment to social justice. *No One Is Illegal Radio*, broadcasting in the rotating 5 p.m. slot on Thursdays, takes on an actively oppositional stance. Its programme Mission Statement reads:

> *No One Is Illegal Radio* highlights the struggles of communities in resistance, with a focus on the self-determination struggles of Indigenous peoples, migrants, refugees and working class people of colour. *No One Is Illegal Radio* is in public confrontation with the Canadian state, organizing to fight against poverty, racism, police brutality, prisons, detentions and deportations, as well as opposing the displacement and genocide of Indigenous peoples on Turtle Island. *No One Is Illegal Radio* actively supports and highlights the organizing of allied groups, including Solidarity Across Borders and the Collectif de résistance antiraciste in Montreal, as well as No One Is Illegal groups across Canada.[37]

CKUT also facilitates activist radio in alternative forms, such as the Dragonroot feminist collective podcast series which foregrounds current struggles around gender representation. Dragonroot also broadcasted on CKUT Tuesday mornings between 8.30 and 9 a.m. in a show produced in collaboration with the Gender Advocacy Centre. Their Mission Statement invites listeners who affiliate with aspects of the struggle as described:

> Dragonroot Media is a feminist media collective that aims to produce anti-oppressive media with a gender focus. Dragonroot values self-determination and representation, and works within an anti-oppressive, anti-racist, and anti-colonial framework. As a feminist media collective, Dragonroot enthusiastically rejects the notion of neutral and objective media, recognizing it as actively supporting established social power dynamics and institutional violences.[38]

CKUT also produces standalone solidarity programmes, which bring in other stations in a shared broadcast around areas of political struggle. On 15 May 2018, CKUT produced and broadcast Radio Free Palestine, an international 24-hour live broadcast in conjunction with other community stations from Palestine, Algeria, Lebanon, Mexico, Morocco, the United States and elsewhere in Canada.[39] The event was broadcast from CKUT and live via Facebook and is archived for subsequent listening on the CKUT website. CKUT's range of affiliation community shows and standalone solidarity programmes provide accessible coverage of issues and experiences for listeners who affiliate with particular political struggles and/or movements. By broadcasting shows via the FM frequency, CKUT enables alternative access to seeking out individual communities or content on online platforms. John Downing (2001) identifies particular characteristics of radical media which fulfil important communicative functions more thoroughly than mainstream media across several areas:

> First, radical alternative media expand the range of information, reflection, and exchange from the often narrow hegemonic limits of mainstream media discourse. . . . Second, they frequently try to be more responsive than mainstream media to the voices and aspirations of the excluded. They often have a close relationship with an ongoing social movement and they fairly spontaneously express views and opinions extruded from mainstream media, or ridiculed in them. . . . Third, radical alternative media do not need to censor themselves in the interests of media moguls [or] entrenched state power. . . . Fourth, their own internal organisation is often much more democratic than hierarchical. (Downing 2001: 44)

In diverse ways, these characteristics, combined or on their own, each facilitates the project of political struggle, whether that project can be defined as raising awareness, providing information and/or facilitating debate towards action or consensus. I draw on Downing's formulation here to draw attention to how community radio shows and stations can put themselves in service to social justice aims located in particular social movements.

Further, the community radio form mobilises its qualities of liveness, intimacy and immediacy in crucial ways for the aims of radicalised communities. Kaddoura feels that presenting the show in a community radio context facilitates more in-depth discussion and topic exploration:

> The advantage is that when we discuss something we can really discuss it. . . . The advantage of community [radio] is that we can structure the show the way we want. And the advantage is that we can go in depth; I think that's the main advantage of community radio, community/campus. That we take a subject. . . . In the mainstream they can say, 'this happened', but you don't understand the context, in what context this comes.
>
> In our show, if we want to take a half hour to go into it, discuss the context, discuss what's behind it, we can do that. Actually that's the point of the show. There's no point to say, 'this is bad', 'Palestinians are not happy'. Okay, let's put that into context. Why are they not happy? Why is this bad? In what context is this bad? It's one of the main things we talk about in the show [*laughs*]: context, context, context. And so the show: yes, it really allows us to go into explaining why this is a problem.

Because we are a community show, I have complete freedom of what I want to talk about, and how, and how long I will go.[40]

As Ayesha Baig puts it:

Community radio's a very intimate thing, I feel. It's more intimate. It's like you feel closer to it as a listener. . . . It gives a closer appeal, an approachable appeal. I feel like, I have access to radio, as a speaker, as a presenter, as a producer. So I'm going to project [that] in the best way possible. In the truest way possible.[41]

Kaddoura elaborates on a similar feeling she has about community radio: 'I prefer these kind of community shows because it feels like more normal people . . . there's a certain organic way of talking that is very charming and seems very real and you don't feel that someone is talking very unnaturally'.[42]

It is the intimacy produced by such 'organic' modes of conversation that—finally—can underscore and cement the appeal and efficacy of activist radio housed by a community station. Through consistently maintaining collective practices of station management, scheduling and decision-making alongside an ongoing commitment to social justice programming, CKUT demonstrates some of the possibilities within community radio infrastructures for not only community-produced radio but for activist radio as well.

CONCLUSION: COMPOUND PRODUCTION AND SOLIDARITY PROGRAMMING

As noted above, CKUT's commitment to social justice emerges from a historical context of a wider political movement for social change in Montreal and Québec dating back to the 1970s—a period which also saw the establishment of Radio Centre-Ville, broadcasting as CINQ FM on 102.3 FM in Montreal. Unlike CKUT's campus-community license, Radio Centre-Ville's community license includes provision for a dedicated multilingual service for Montreal's diverse ethnic and linguistic communities. Since its inception, the station has broadcast regular programmes in first seven, later eight languages, produced by as many as three hundred volunteers (Foy et al./Radio Centre-Ville 1992). As defined by its practitioners in 1992, '[t]he station acts as a catalyst between communities and the entire social and cultural fabric, which, in turn, has been enriched by the presence of the communities'. The first two principles of Radio Centre-Ville's mandate at time of writing are:

To broadcast radio programs aimed at Montreal residents, to provide them with a clearer understanding of issues related to their living conditions, the problems they experience, and the expansion of their own cultural awareness.

To enable citizens, especially those usually ignored by other news media, to express themselves over the airwaves.[43]

In reference to its inclusive mandate, Radio Centre-Ville represented their overarching project as one of political and cultural exchange with the wider Montreal community: ‘First and foremost a centre for radio broadcasting, Radio Centre-Ville is also a centre for meeting and exchanging ideas and social and cultural experiences’ (Foy et al./Radio Centre-Ville 1992: 49). As at CKUT, this commitment to inclusivity also manifested in Radio Centre-Ville’s working practices, which have been shaped by the strength of its volunteer base. Management structures at the station were built around collective participation of its autonomous volunteer production teams. Echoing a Marxist approach to control of the means of cultural production, Radio Centre-Ville foregrounded collective volunteer participation at management and decision-making levels as well as in programme production: ‘Good intentions are not enough to ensure that the target population and groups participate in the production of radio broadcasts. These groups must take control of the means of communication by participating in the management and decision-making process. In this way, they manage the complete production process of the station’ (Foy et al./Radio Centre-Ville 1992: 54).

As we have seen, CKUT’s station practices work along similar lines. Unsurprisingly, given their shared commitment to social justice and active promotion of diverse and equitable cultural and linguistic programming, CKUT and Radio Centre-Ville have a longstanding working relationship and have developed collective approaches in tandem. CKUT provides some programming originally conceived of by Radio Centre-Ville, such as prison radio and standalone solidarity programmes.

In the spring of 2017, Radio Centre-Ville underwent a fundamental crisis in which a central tenet of community radio, to provide community practitioners with regular and reliable broadcast space to represent their communities, was effectively repealed for many groups served by the station. Radio Centre-Ville was broadcasting shows in eight languages in 2017, including Arabic, Chinese, English, French, Greek, Haitian Creole, Portuguese and Spanish. However, reportedly due to a substantial loss of key grant funding, the station somewhat abruptly changed its scheduling practice and started to require that each show in its schedule raise funds independently. Several community shows were subsequently removed from the schedule and the airtime sold.[44] The scheduling changes were part of wider changes brought in to cut costs at the station: paid staff were made redundant and membership fees increased substantially, along with other changes. This compound decision to require each show to pay a fee and to lease out airtime deepened the crisis. The crisis was simultaneously fiscal, legal and political, and the divide across Radio Centre-Ville’s groups reportedly became deep and bitter. Selling airtime goes against the principles of community radio in general and came as a shock to the reportedly abruptly removed community groups at Radio Centre-Ville in particular. Further, as decision-making at the station had always been on a collectively agreed upon basis, community members forced off the air felt doubly betrayed. Station manager Wannex Lalanne Zéphyr argued the measures were necessary to keep Radio Centre-Ville from going completely under. Dissenting community groups who felt forced off the air argued that if the station was to become a 'top-down commercial enterprise' it shouldn't continue to exist. The two dissenting groups held two separate assemblies to vote on future actions in spring 2017, but neither side recognised the other in the voting process.[45]

In May 2017, CKUT staff came to the aid of some Radio Centre-Ville practitioners by hosting and broadcasting their programmes from CKUT when they were denied access to their established slot on the Radio Centre-Ville schedule—building on their established reciprocal relationship in community radio practice. Tamara Filyavich is Arts and Culture coordinator at CKUT and was involved in helping some Radio Centre-Ville practitioners broadcast at CKUT, including two Haitian community teams, two Francophone teams and a Spanish-language team. They initially met with these community groups to organise a single day of programming but soon determined a longer-term solution was needed for these groups to continue to represent their communities. At the same time, CKUT found itself with several schedule slots to fill due to staff changes affecting its news talk coverage.[46] Thus the Radio Centre-Ville groups began broadcasting live between 5 and 6 p.m. Mondays through Fridays in the desirable news slot as the news director had just left the station. As Filyavich confirms:

> It was the perfect fit: we needed the time [filled on air]; they needed a platform. . . . We trained them on our equipment and the particularities of doing radio on CKUT but we didn't dictate any of the content and they essentially continued doing some of the programming they were doing . . . because their content fit in with their mandate anyway.[47]

CKUT hosted several groups between May and November 2017, at which time most of the practitioners returned to Radio Centre-Ville following a court decision in their favour. Filyavich points out that CKUT was never asked to take a position during this bitter dispute and never did so:

> We were never asked to take that position; we were just asked to make space for a sister radio station essentially, that was, the same way as us, one of the first third language programming stations in Quebec. . . .

> Plus, some of our members have individually mobilised them, but again not as CKUT. Which is cool, because people should have that freedom, for sure.[48]

At time of writing, while Radio Centre-Ville remains on air, the dispute has not been fully resolved.

It is crucial to recognise that the crisis faced and experienced by Radio Centre-Ville's management and practitioners could happen at almost any community radio station. Radio Centre-Ville was plunged into sudden structural precarity when substantial financial resources that the station could previously rely on were removed. In turn, the station had to struggle to sustain itself. The ways in which Radio Centre-Ville attempted to stay on air remain contested but their position of resource precarity will be a familiar scenario to anyone involved in community radio. The responses by Radio Centre-Ville practitioners and CKUT staff exemplify a particular set of working relationships, principles and practices specific to this moment in Montreal, yet the collective support provided can also be taken as an example of good practice in a crisis situation which almost any community station could find itself negotiating.

NOTES

1. Canada's other community radio licensing provision includes dedicated provision for Indigenous communities and multilingual programming, as with Montreal's Radio Centre-Ville.
2. Louise Burns in interview with Joni Sadler, 6 June 2017.
3. Joni Sadler, Burns/Sadler interview, 6 June 2017.
4. Burns, Burns/Sadler interview.
5. Available via https://ckut.ca/sites/ckut.ca/files/the_blue_book_2016-17.pdf (accessed 24 April 2018).
6. See Kruse 2003 and Wall 2007 for in-depth discussion of alternative and 'indie' music scenes in radio.
7. Schedule as of May 2018, from https://www.ckut.ca/en/content/ckut-903-fm (accessed 10 May 2018). Description of *Prison Radio* from show webpage, https://prisonradioshow.wordpress.com/about/ (accessed 6 July 2018).
8. Burns, Burns/Sadler interview.
9. Sadler, Burns/Sadler interview.
10. Sadler, Burns/Sadler interview.
11. Burns, Burns/Sadler interview.
12. Sadler, Burns/Sadler interview.
13. From the CKUT schedule, https://secure.ckut.ca/cgi-bin/ckut-gridnew.pl (accessed 12 June 2018).
14. Burns, Burns/Sadler interview.

15. Sawssan Kaddoura sometimes coproduces the show with Katherine Garven (from interview with Kaddoura, 21 June 2017).

16. From interviews with Sawssan Kaddoura, 21 June 2017, and Ayesha Baig, 27 June 2017.

17. Interview with Ayesha Baig, 27 June 2017.

18. Baig interview.

19. Interview with Kaddoura, 21 June 2017.

20. Burns, Burns/Sadler interview, 6 June 2017.

21. Baig interview.

22. Baig interview.

23. Baig interview.

24. Burns, Burns/Sadler interview.

25. Baig interview.

26. Baig interview.

27. Baig interview.

28. From https://www.ckut.ca/en/content/under-olive-tree (accessed 12 April 2018).

29. Kaddoura interview.

30. Kaddoura interview.

31. Kaddoura interview.

32. Kaddoura interview.

33. Kaddoura interview.

34. Kaddoura interview.

35. Kaddoura interview.

36. Kaddoura interview.

37. Statement from the show homepage at http://ckut.ca/en/content/no-one-illegal-radio (accessed 15 June 2018).

38. From https://ckut.ca/en/podcasts; see also http://www.dragonrootmedia.com/radio-archives (both accessed 6 July 2018).

39. Details and information found at https://www.ckut.ca/en/content/radio-free-palestine-2018 (accessed 29 May 2018).

40. Kaddoura interview.

41. Baig interview.

42. Kaddoura interview.

43. See Radio Centre-Ville's stated mandate at http://radiocentreville.com/wp/en/information/the-mandate/ (accessed 6 July 2018).

44. See Faguy, S., 'Radio Centre-Ville at Centre of Power Struggle', *Montreal Gazette*, 31 March 2017, http://montrealgazette.com/news/local-news/radio-centre-ville-at-centre-of-power-struggle (accessed 11 May 2018). See also https://blog.fagstein.com/2017/05/28/radio-centre-ville-crisis/ (accessed 6 July 2018).

45. From https://blog.fagstein.com/2017/05/28/radio-centre-ville-crisis/ (accessed 6 July 2018).

46. Interview with Tamara Filyavich, 21 June 2018.

47. Filyavich interview.

48. Filyavich interview.

Chapter Four

Producing Alternative Publics in Toronto Campus-Community Radio

Community radio is regularly associated with alternativism, itself another term—like community—with many connotations. Stephen Barnard (2000) confirms the diagnosis of community radio as alternative radio, suggesting that as a form community radio functions as an alternative response to the mainstream—as explored in the previous chapter. Given the fluidity of alternativism as a descriptor, this chapter focuses on ways in which community programming in Toronto seeks to represent alternative communities in the city. Considered as the commercial capital of Canada, Toronto combines attributes of a globalised capital with a highly functioning infrastructure with a plenitude of green spaces. Toronto is regularly represented as a success story of diversity and is home to established Chinese, Italian and Portuguese neighbourhoods alongside multiple other ethnic and linguistic communities. At the same time, property values in Toronto have soared and it is today extremely expensive to buy a home there. In this chapter I explore how marginalised communities are represented on air in a particularly contradictory city with a vivid sense of its own diversity and progressiveness. In this exploration I identify ways in which 'community' is considered in the community radio programmes under discussion, and also examine how communities draw on and produce a collective sense of identity and reinforce and reproduce this on air. Here I am considering production practices as 'deployments of community', drawing on Miranda Joseph's (2002) recognition of the ways in which 'community', when framed as an 'alternative', can be considered both as *other* to mainstream broadcast representations of the communities involved and as a necessary adjunct sector of that same wider broadcasting environment. In this consideration, I build on my discussion of the societal and political functions of 'community' first developed in chapter 1 and expand on these to explore more thoroughly the representative functions of selected

Toronto community programmes as such deployments of 'alternativism'. In particular, I examine the ways in which production practices in community radio function to reproduce, and thus legitimate, apparently alternative or absent subjectivities, which then become a symbolic as well as labour resource for the station and its representative project(s).[1]

This exploration begins with the understanding that reproductions of apparently alternative communities can be understood as *complementary* to the overarching mechanisms of capitalism; as located in the same Venn diagram, rather than representing a separate space of articulation. As argued in chapter 1, a persistent conception of 'community' as a de facto embodiment of a localised, positively collective alternative helps to explain the Romantic attributes the notion is saturated with, but is ultimately unhelpful in its simplicity and sentimentality. Instead *community* is more usefully recognised in terms of its discursive and practice-centred mobilisation, following David Harvey (2001) and Joseph's excavation of these in particular. Thus, 'Deployments of community, both verbal invocations and practice, are conditioned by a larger discourse of community, a pervasive way of thinking and doing community . . . [t]hat sets the terms in which we might ask questions, and shapes what we can see and do and even who we are' (Joseph 2002: 1).

In this formation, 'community' is mobilised as a mechanism to frame and think through community practices and identity (re)production. As particularly discussed in the previous chapter, such deployments of 'community' are often embedded in community station guidelines and remit narratives, informing inclusive station practice while simultaneously delineating the parameters of that practice. Chapter 3's discussion demonstrated how CKUT deploys discourses of community in a persistently collective set of approaches, for example. This 'discourse of community' also acts to position and reinforce an understanding of 'community' as a 'defining other' (Joseph 2002: 1), marking out and reifying a site of difference which then reinforces the smooth running of capitalist society. Given capitalist tendencies of appropriation, understandings of community must necessarily instead be complicated as we identify and differentiate between instances where 'the practice of community might offer effective resistance and where it might be an unredeemable site of co-optation, hegemony, and oppressive reiteration of norms' as Joseph insists, drawing on Eve Kosofsky Sedgwick (Joseph 2002: xxv). This recognition enables a determination of both the overarching project of a given community station, and the capacities for resistance or co-optation of individually produced programmes of that station. This complex understanding and mobilisation of community as an alternative social form, as well as a media one, is all the more necessary as community radio itself has long been associated with alternativism, so as a form already embodies these attributes and mythologies which come complete with positive and potentially uncritical connotations of otherness. This analytical lens is also necessary to unpack the actual capacities and possibilities for resistance offered by such alternativism where these can be deployed, given capitalism's tendency to co-opt activities and texts considered as 'alternative' for containment and commodification.

In a nutshell, community as it functions in ways identified by Harvey and Joseph and others has the capacity to usefully embody a productive otherness: consequently, community radio practices and programme content inhabit an *other* broadcast space. Building on prior chapters' explorations of accented, activist and other community-led strategies for community self-representation, in this chapter I aim to identify further strategies of community-building enabled by these alternative spaces of articulation and expression and explore how these are deployed to encourage and reinforce on-air spaces of affiliation across self-identified alternative communities. I want in the process to explore how a sense of community identity is reinforced and sustained through the production of structures of feeling which function to connect community members through expressed shared experiences reflecting the particularities of their community affiliation.

DEFINING AND SCHEDULING DIVERSITY ACROSS COMMERCIAL AND CAMPUS-COMMUNITY STATIONS

Radio representations of Toronto's considerable and long-established diversity—defined discursively through cultural, ethnic, and linguistic affiliations—can be heard in selected public service broadcasting (PSB) coverage of the city's multiple communities alongside standalone community programming broadcast on dedicated multicultural and campus-community stations. These include stations self-defined as multiethnic, the two most established of which are the CHIN radio group and CIRV FM. At the time of my 2016 research, LGBTQ communities are represented in Toronto radio in part by the dedicated commercial station CIRR FM, which brands itself as Proud FM. In addition to these standalone stations, three university stations in Toronto previously operated on campus-community licenses: CKLN FM at Ryerson, CIUT FM at University of Toronto and CHRY FM at York University. CKLN FM has closed down, while CHRY FM rebranded as VIBE FM, with a focus on music superseding its prior community-programming focus. CIUT, however, continues to broadcast dedicated community programming in addition to student-produced shows. When mapping Toronto's radio stations and provision of community programming, I began to see distinct differences between radio representations of minority communities defined primarily by ethnic and linguistic affiliation and communities identifying as otherwise 'alternative'. Toronto's historical ethnic and linguistic diversity has produced two FM radio stations dedicated to multiethnic programming: CHIN FM (also incorporating an AM frequency) and CIRV FM. I suggest there is a structure and hierarchy in Toronto's alternative radio ecology in which CHIN and CIRV represent 'official' diversity programming, continuing and reinforcing a multicultural project of containing and reinforcing hierarchies of difference among minority communities. In this context, CIUT then takes up representation of 'other' communities considered as marginalised or alternative in its broadcast schedule—particularly following the demise of Toronto's other two university radio stations, which deepened the absence of alternative programming from the city's airwaves. While both CHIN and CIRV dedicate their broadcast space entirely to Toronto's diverse ethnic and cultural communities, their mandated focus on multiethnic representation serves to simultaneously reinforce these communities as minority or other communities—an established multicultural strategy of othering (see Gordon and Newfield 1996, in particular). To further contextualise this diversity programming, Toronto's Canadian Broadcasting Corporation (CBC) stations (Radio 1, Radio 2) do incorporate some First Nations programming, but only up to a point. There is some diversity in presenters but the majority are primarily identifiable as Anglo Canadian.

Figure 4.1. CHIN radio studios, Toronto, June 2016. Photo supplied by the author.

CHIN FM has been broadcasting for over thirty years from spacious studios on College Street in central Toronto which also incorporates studios for CHIN radio group's AM frequency. The CHIN radio group also incorporates a third multiethnic station broadcasting from Ottawa. The main communities represented across the CHIN schedules (with separate schedules for FM and AM frequencies), self-identified by national and linguistic affiliation, include Cantonese, Italian, Mandarin, Portuguese, South Asian, Spanish, Tamil; while shorter programme slots are dedicated to Bosnian, Croatian, Greek, Jewish, Polish, Russian and Ukrainian shows. These programmes, many of which have been running for a decade or more, are primarily presented in the first language of the target communities. Community groups are allocated programme slots in the schedule relative to their popularity and the needs of their community target audiences in an established schedule. Station manager Theresa Lombardi describes the process: 'We like to say all our programming is prime time to the community it serves. So it may not be on a Monday to Friday . . . but that community knows to tune in on a Saturday and that's prime time to them, because they make time to listen'.[2]

Lombardi stresses that CHIN FM operates as a commercial station rather than a nonprofit, but at the same time describes a model of community programming typical of community radio: established volunteer programme producer-presenters whose primary motivation is passion for the work and who are trusted representatives of their respective communities, in part through their visibility as radio show hosts. Lombardi describes show hosts as particularly dedicated:

> It's a heartwarming tradition to serve your community, and all our presenters are dedicated to their communities. . . . Most of them, it's not a job to them, it doesn't pay them, it just keeps them in contact with their communities. . . . So, long-term, they [producers-presenters] come, they stay . . .
>
> They're known in their communities, so they come with a mind to serving their own community. . . . It's a way to make the whole community inclusive.[3]

The 'heartwarming' quality described here is characteristic of the romantic mythology around community noted by Miranda Joseph (2002); this quality as described here, along with the performed passion of these volunteer practitioners for their work, also serves as a rationale to the tendency in community-focused programming to perform community outreach work on an unpaid basis. As articulated throughout this book, the volunteer nature of individual programme work is a testament to the passion felt by show hosts for their programme and its cultural and/or political remit as they perceive it. I want with this chapter to explore further how this labour, conceived collectively as driving community programming, is directed towards representation of an alternative set of subjectivities. The volunteer nature of this work means that in community radio's political economy the programme producer is usually also the main presenter, as well as researcher, and usually promotes the show on social media as well. I suggest that the alternative articulations of identity and experience which community radio enables and produces are not only expressed on air but also circulate in community station programming and scheduling. The broader structural function of nonprofit community radio stations means that they function dialectically with mainstream broadcast institutions, picking up the slack in cultural, minority community representation where—in Toronto—public service and commercial services do so only in limited ways. Thus, as Joseph argues, 'Non-profits mobilise for capital—but not only for capital—subjectivities that are not (only) of it—are in fact its absent center—and yet set it in motion' (Joseph 2002: 118). In short, community radio and community programming, in facilitating representation of minority communities, enable articulation of these communities—drawing on volunteer labour to do so—where this is 'absent' elsewhere in Toronto's radio ecology.

University of Toronto station CIUT features a more pronounced diversity in its multifaceted schedule. CIUT's schedule contains a mixture of arts and cultural programmes; programming from ethnic and linguistic communities, including one First Nations show, *Indigenous Waves*; LGBTQ and sex-positive programmes; issue-based programmes (for example, *Green Majority* focuses on environmental issues) and a range of current events and music programming which includes student-produced programmes. CIUT's particularly diverse schedule is facilitated by the conditions of its broadcast license, which station manager Ken Stowar describes as 'very specific'. Like Montreal's CKUT, CIUT operates under a campus-community license, which enables and explicitly calls for wide representation of Toronto's multiple community groups. CIUT has been broadcasting for over thirty years; its offices and production and broadcast studios are located across two floors in the one-hundred-year-old Hart House, a striking building in the Gothic Revival style and located near the centre of the University of Toronto campus. The diversity of the schedule reproduces diverse community groups and community interests in the city, incorporating communities self-defined by ethnicity and/or language, but also communities which share cultural and/or political affiliations and interests. CIUT's schedule represents a complex combination of communities, representing an inclusive understanding of 'community' similar to that of CKUT. CIUT's mission statement's commitment to inclusivity is summarised as follows: 'Based on our mission statement, we provide an alternative to mainstream media, provide community access to the airwaves, and offer programming that reflects the diverse elements of the communities we serve'.[4] Stowar describes how the station's remit is then interpreted in their scheduling: 'What's embodied in that is we *must* be different than the private sector, the public—meaning CBC [Canadian Broadcasting Corporation]—and any other type of licensed station. However, at the same time, I always let people know, no one radio station can literally address every single community. Impossible'.[5]

As part of this mandated diversity, CIUT provides schedule space for communities otherwise underrepresented by other Toronto radio provision. The diverse accents heard on many of CIUT's community-led programmes are characteristic of the station's project of inclusivity. Stowar provides context for this initially by citing the station's broadcasted range of accents as evidence of CIUT's difference from mainstream or commercial radio in Toronto: 'The private sector over here, you can't have an accent.' However, as argued throughout this book, to be accessible to members of their target communities, community programmes made by and for their target groups are often presented and produced by members of those groups—with the accent functioning to perform community affiliation, which in turn serves to draw in listeners from that community, in particular.[6] The role of the accent is considered inherent to this programming by Stowar:

> I get a lot of people coming to me from potential communities. . . . And I assure them that, as long as you can enunciate, there is no barrier here when it comes to accents or dialects, because it represents where we live. . . . And I do state

Figure 4.2. CIUT studios, University of Toronto, June 2016. Photo supplied by the author.

> that boldly, because individually I know they're a little shy, and reserved and hesitant, so . . . Let them know that there's opportunity, and there's no barrier to having an accent. There's all kinds of people here who have accents. Where we live!
>
> Play to the face of the programme, play to the face of your community.[7]

As with community radio programmes elsewhere, CIUT's shows are almost entirely produced and presented by volunteers. The exceptions to this at CIUT are occasional instances (as with the producer-presenter of CIUT's *Rainbow Country* show, Mark Tara) in which the show host also works in another capacity (production, scheduling or other station work) for the station. As Stowar notes, 'It costs money to volunteer', but it also costs time and energy. Echoing Lombardi's observation that for established volunteers, producing and presenting their shows isn't 'a job to them', Stowar describes the degree of commitment of CIUT volunteers:

> First and foremost, it's passion. The passion is there. . . . They are engaged on a day to day basis, outside of this radio station, and they bring it in here. You know, when it comes to music that's simpler to explain. There's a lot of people that are attached to, immersed in, particular scenes: the punk scene, the psych scene, and they bring with them a reputation before them. Sometimes. Or that reputation develops here.[8]

Each community show manifests a particular specificity in how community ideas, perspectives, problems and negotiations are expressed.

> If you're going to do a show for a specific community, be inclusive. Because people outside wish to know what your concerns are, what your perspectives are and so on. . . . You can't give everyone the same message in the same way. Everyone is different. You have to know how to do it.[9]

As at CKUT, the diversity of programming at CIUT incorporates shows by and for communities of interest and of affiliation. *Bring Ya Eh Game*, the title a sly play on perceptions of a specifically Ontario pronunciation of 'eh' to replace 'a', has showcased Canadian hip-hop since June 2015. Broadcasting live on Monday nights from 10 p.m. to midnight, the show is presented and produced by Justin Gunderson, aka 'Ya Favourite Lightskin', also a hip-hop artist who is able therefore to speak from inside the scene with an insider's knowledge. The show has developed a loyal following, encouraged by Gunderson's comprehensive provision of show content across social media platforms. Despite the increasing popularity of online platforms as a place to source new music, however, Gunderson argues for the continuing importance of community radio as a platform:

> I think it [radio], especially community radio, is important because it's always the voice of the unheard. I mean, you have social media and everyone can go online and people do a lot of online streaming these days and a lot of people consider radio an online thing now, but I think it's important to still have that [analogue radio].
>
> It's like you have CDs and MP3 but vinyl's always something that's going to be classic, and I think radio is that too. And if it fades away then music is kind of lost, because you don't have that place to go to, to hear the local stuff that nobody's playing.[10]

Gunderson's commitment to promote local hip-hop and urban music is further borne out in the production of four CD volumes featuring local artists in a series called *The Eh List*, expanding promotional possibilities further. *Bring Ya Eh Game* regularly incorporates returning guests into the studio to contribute to discussion about new releases and hip-hop and urban music events in Toronto and elsewhere in Canada.

Another CIUT programme, the long-running *The Green Majority,* focuses on environmental issues and again has a loyal following, both in Toronto and nationally, as the show is syndicated across twenty-two radio stations.[11] Broadcasting live at CIUT on Friday mornings between 11 a.m. and 12 noon, and also producing podcasts accessible via the *Green Majority* website, the show has been produced and presented for over ten years by Daryn Caister. Caister envisions their target listeners very clearly, as attuned to environmental issues and seeking a sense of community:

> Essentially what the point of the show is, there's a lot of folks out there who are very interested in these issues, who are very concerned about these issues, they want to learn and/or they want a sense of community with other people that are also interested. But they're not necessarily people who, and I think this is most of our audience, people who can't really—either don't have the time or choose not to actively engage with civil society on these issues, so people who don't have time to be a volunteer or an activist or maybe that's not how they want to be involved, so we design our show specifically for interested nonactivist parties.
>
> So we're basically trying to create a conduit for the people who are our peers, people who are super plugged into these things, in many cases people who work at not-for-profits, and communicate their expertise to people for whom the show or podcast is probably their only exposure to these issues, aside from maybe reading news articles occasionally. So it's educating an audience that are already interested but we don't assume that they know very much, and we're kind of their connection to the movement.[12]

Both *Bring Ya Eh Game* and *The Green Majority* have an established and well-produced online presence; each has a website which links in turn to content across further platforms including Instagram, Twitter and YouTube.[13] This wider online reach both extends platforms for accessing show-related content and builds trust in audiences who prefer and rely on these platforms instead of listening to radio via the FM signal and/or in real time. Multiplatform reach increasingly typifies the practice of many community shows, as seen in chapter 2's discussion of *The Morning Drive*, facilitating greater access which in turn expands a show's listenership. Caister's clear sense of his listening public demonstrates a tangible knowledge of *The Green Majority*'s influence, and in turn, reinforces a self-imposed mandate of accountability characteristic of activist radio, which comes through as he describes the importance of the programme's project:

> I think we provide a vital service. And I think a lot of people would not understand what's actually going on without our show. It's not because I'm egotistical, it's not because I think that I'm the only person who could do what I'm doing, but we're certainly the only people who *are* doing it.
>
> There are certain points of view and people who get silenced that without shows like ours and other mediums like us, who simply would be never heard of otherwise, and they happen to have incredibly important and valuable things to say.
>
> I'm going to sound cheesy for a second, but it's sort of a with a great power comes great responsibility thing. Somebody's going to be on that airtime, maybe they'll want to do a music show, I think that what we're doing is really bloody important, so I feel a bit of responsibility around my show. I feel like, if we don't do it, no one else is going to. I believe that to be true.[14]

The range of detailed reporting provided on *The Green Majority* website (including links to the live programme and to standalone podcasts) supports Caister's description of the programme's self-imposed watchdog remit and critical coverage of environmental issues and crises. Caister explains that he is given significant freedom at CIUT to produce his show, and experiences complete support from Stowar as manager: 'He's never ever once in almost ten years told us what I should or shouldn't talk about on our show. He's just kind of looking for people who I think are just committed to the idea of community radio and I think that's the people he tends to really support'.[15]

While *The Green Majority* covers environmental issues around Canada and in the United States, Caister suggests that Toronto embodies particular contradictions which further underscores the importance of an on-air space for articulating progressive perspectives on green and environmental concerns.

> It's a really modern city and super accepting and [known for being] the most diverse city . . . but at the same time the reason it's a big city is because it's full of corporations and lots of people who work there and it's Canada's Wall Street and all that sort of stuff.

> I think that's part of why we do the show, because . . . it's sort of an ultimate clashing of worlds here and so that's why it can be such an exciting place to be sometimes, because you'll have a whole bunch of big corporations like Barrick Gold being headquartered here and doing all sorts of terrible things and us being able to have an activist meeting and literally go and bang down their door and say 'what are you doing about it?'[16]

Through actively foregrounding critical perspectives on environmental issues from an established platform, *The Green Majority* combines provision of a discussion forum for acute environmental community issues with crucial information provision. As in chapter 3's discussion of ways in which activist radio encourages participation through political and community affiliations, *The Green Majority*'s remit and project pulls together a community coalescing around shared concerns while representing alternative coverage of environmental issues to that circulated by corporate interests held in Toronto and elsewhere.

QUEER CULTURE BUILDING ON AIR

I argue throughout this book that a shared sense of community is produced and reinforced by community-focused programmes, produced and broadcast via dedicated on-air spaces to facilitate community-led articulations of identity and situatedness. In Toronto, Proud FM represents LGBTQ communities; however, as of summer 2016 its schedule was sparse and incomplete, with swathes of airtime unoccupied. Ryerson University's station, newly branded as CJRU AM in 2016 and calling itself The Scope, broadcasts *Perfectly Queer* on Saturday nights between 9 and 10 p.m. CIUT's schedule incorporates spaces for programming which can be read as queer. In 2016, two CIUT shows broadcasted to and for Toronto's LGBTQ and sex communities: *Sex City*, broadcasting for over a decade, and *Rainbow Country*, which began broadcasting in spring 2016. *Sex City* foregrounds a wide range of topics related to sexuality, including sexual health information and education, debates over current issues in the city's sex communities and coverage of related community events. *Sex City* is produced and presented in rotation by a team who alternate shows: sometimes two hosts will present three items in the scheduled hour while at other times one of the hosts produces and presents the whole hour.[17] For the *Sex City* production team, it is important that the show's content is accessible; established producer-presenter Jon Pressick explains and unpacks how they approach this:

> We do aim to appeal to a general audience . . . we very much mean to *not* speak from the 'inside'. . . . We present topics: 'oh, you're launching a trans-clothing

line. Can you explain *why* we need a trans clothing line? Can you explain *who* needs a trans clothing line?'

We want to make the show accessible. So our primary audience is everyone. Our secondary audience is the sex community, so that people can understand what's going on in the community, people can see the events that are happening in the community. They can be an entrance point for the general audience; they can also be a meeting point for people who are in the community. And then it breaks down into all of the various, countless sex communities, queer communities, gender communities, which then you splinter down into so many different branches of a tree, it's like . . . there's so many. Even from when I started on the show to now, it's a massive difference [*laughs*] into who you can address and what topics you can find.[18]

By consciously working to not speak 'from the inside', Pressick and the other producers aim to keep the show's language and points of reference accessible for and intelligible to new as well as familiar listeners. Other producer-presenters for *Sex City* are Louise Bak, Dorianne Emmerton, Sophie Delancey and Cordelia Huxtable[19]; having multiple practitioners on the programme serves several functions, ensuring an equitable division of labour but also enabling provision of multiple subjectivities and perspectives, encouraging in turn multiple points of identification for listeners. Pressick says each of the show's hosts have distinct styles of presenting, and connects this to the programme's capacity for representing Toronto's diverse sex-affiliated communities:

Because we're weekly, because we have such diverse hosts, I don't think we're leaving any communities out for any great lengths of time. Toronto has a strong sex community. But Toronto is not huge. So we can almost run out of content on occasion, it sort of seems. But we can cover anything. We can do three different subjects during the show, or we can theme the show.[20]

Rainbow Country began broadcasting in spring 2016 and is produced and presented by Mark Tara, who sees his show as dedicated to the gay community with an inclusive remit. Tara believes his programme is ideally situated to host debates on issues related to the gay community, as well as reaching a wider audience interested in these ideas and issues. Tara describes the show's project and audience as having an inclusive remit for the LGBTQ community:

It's a gay radio show, it's a LGBT radio show absolutely. Yeah, so I'm reaching out to all aspects of the gay community. Whether they're people of colour, Asian individuals, Native Canadian two-spirited individuals, lesbians, the leather community. Because I think a lot of times certain people are represented, and over-represented, and then you don't hear a lot from say, coloured gay individuals. Because I think a lot of it comes from upbringing, religious upbringing.

And a lot of people don't want to be found out. They don't want to be cut off from their families, because of upbringing . . .

So for my show, I'm really looking for that untold story, that untold aspect, because that's really my underlying mission. Because it's an LGBT show, but beyond that I want to really be representative of people that really haven't had a voice, up until this point.[21]

Tara's emphasis on utilising the broadcast space provided in *Rainbow Country* as a site for individual stories facilitates articulations of diverse experiences which LGBTQ allies and affiliated listeners can relate to.[22] Both *Sex City* and *Rainbow Country* provide dedicated broadcast space for the expression of diverse experiences of queer subjectivity. Tara describes *Rainbow Country*'s focus on LGBTQ communities as addressing a gap in Ontario's radio ecology, expanding on this when I reference Toronto self-defined gay station, Proud FM, and identifying what he sees as key issues for LGBTQ people and communities:

It's a gay station, but no one has a show like mine. . . . So there is a lack, and there has been a lack in southern Ontario. There's no other gay programme that I'm aware of. In terms of a radio programme . . . to my knowledge? My show is unique.

. . . I think too that the issues of today are different to the issues of yesterday. Back then, the gay community was fighting for the right to adopt, or the right to marry, or the right to just be who you want to be. And I think a lot of those things have been checked off the box, checked off the list. To a certain degree. But because they're checked off doesn't mean that there's acceptance. You can't legalise etiquette. You can't legalise . . . acceptance. You can't legalise lack of racism or lack of homophobia. You can't do that shit. Excuse my language.

So is there a lack? To me there is, because I don't hear anything that I want to listen to.[23]

Before the inception of *Rainbow Country*, Tara was already working at CIUT in a production role; he also previously contributed to *Sex City*. He pitched the idea for *Rainbow Country* to station manager Ken Stowar, insisting there was a real need for a LGBTQ focused show: 'I made the valid point that a show like this is still needed, in 2016. Especially since what happened in Orlando. I said we need a show like this. So he obviously saw that there was a need, and that it would be supported by the community, and so here we are'.[23]

As a recording artist himself, Tara additionally felt there should be a broadcast space to promote work from LGBTQ musicians.

It's interesting to me because I play music from LGBT artists, and I personally don't know of a radio station that's playing gay artists from around the world—and I am. And I feel very proud of that, because as a gay artist myself,

> a [place] like this didn't exist. So I wanted to create a stage, and a platform and opportunities, that I feel are lacking, because I wasn't able to tap into something like that because it didn't exist at the time, so therefore there's a need for that, however niche it might be, there's still a need for that.[25]

However, Tara doesn't necessarily see promotion of LGBTQ artists as the central impetus of the show. Instead he hopes the show will provide a forum for discussion and debate of key and ongoing issues for Toronto's LGBTQ communities. Tara suggests that music promotion is fine, 'but it's not really what I want to use the show for, which is to raise issues and to be controversial and to say, to talk about issues that people don't want to talk about, but talk about in a respectful way'.[26]

Tara calls *Rainbow Country* a 'Gay Radio Show' (on air and also on social media) and introduces each show with the following statement:

> No matter who you are, no matter what you do, everyone has a story to tell. And right here, on this LGBT show, is where it all starts. We'll lift our voices, share our stories and we'll raise our pride. After all, we've come a long way since 1969 and the Stonewall riots, to the legalisation of same-sex marriage, to the recognition of trans equality. I'm Mark Tara. Join me on this journey through Rainbow Country. (*Rainbow Country*, 7 September 2016)

Broadcast live on air from the CIUT studios between midnight and 1 a.m. on Wednesdays, *Rainbow Country* combines music and talk in a balanced and structured way: 'There is structure. Because we do have certain things that we can and can't do. So my show is a blend of talk and music. To me, CIUT seems more commercial, without the commerciality. It's not like the Wild West. . . . We're obviously freer to talk about certain things, certain issues or what have you'.[27]

That freedom from a preset, station-mandated running order (typical of both PSB and commercial mainstream radio), combined with the midnight time slot, potentially facilitates more free and frank discussion and debate. Tara explains that he also considers pacing to be central to retaining audience interest, particularly of younger listeners who may be accustomed to a faster paced presentation style. He feels that community radio's lack of a rigidly enforced running order dependent on making time for preplanned advertisements provides a greater freedom for ensuring a programme's flow. *Rainbow Country*'s liveness combined with its late-night slot (with further confessional possibilities) means that it is especially well situated for on-air discussion of current issues for LGBTQ community members.

The show's liveness is representative of CIUT's programming, as most of the station's shows are broadcast live (which is also characteristic of community radio practice more broadly). Stowar explains that this is a deliberate policy and practice of the station.

> For me, it is important that the majority are done live. You need to be there for your audience. Prerecord, you lose something. You have no one there to respond, or interact with the host, or guest; it's not there. So I do my utmost to make sure the majority of programming goes out live. Live: there's an energy, there's a spontaneity, there's things you can't put into play with prerecord.[28]

Topics covered on *Rainbow Country* in summer and autumn 2016 included interviews with Toronto's LGBTQ rugby team (31 August) and Toronto's Bisexual Network (21 September), and with LGBTQ-identified people about their individual journeys, incorporating music artists, drag artists, gay estate agents (19 October) and other members of Toronto's LGBTQ community (17 August; 24 August; 5 October; 12 October). The 7 September episode featured comprehensive coverage of the Pride Toronto Town Hall Meeting in a dedicated three-hour show, in which the meeting held the week before on 30 August 2016 was recorded for broadcast and only lightly edited to include station IDs throughout. At the start of this show, Tara describes the meeting in a balanced way, and foregrounds the importance of the meeting as a forum for Toronto's LGBTQ communities:

> It was heated at times, and you'll hear that. . . . People spoke their minds, so they should, and I think it will be up to you, the listening audience, what you take from this, if you think it was a fair representation, if you think the crowd was potentially skewed to one side or the other. I think that's an interesting discussion to have after you hear this. . . . It's going to be a long episode. I'll be here with you, but I hope if you can't stay for the whole thing, I hope you'll download the episode . . . because I think it's worth hearing what people have to say, where they're coming from. (*Rainbow Country*, 7 September 2016)

In providing comprehensive coverage of the Pride Town Hall Meeting on *Rainbow Country,* Tara avails of radio's historical potential to function as a broadcaster of significant events for political and social education and information (see Scannell 1991). The three-hour slot provided for this particular *Rainbow Country* by CIUT indicates the station's support for broadcasting the meeting and the views of those groups and communities represented. The meeting itself included contributions from multiple LGBTQ representatives from groups identified as LGBTQ black, First Nations, Muslim, and other ethnically and culturally defined communities. The meeting included contributions from the floor following a conflict which emerged at the 2016 Toronto Pride parade when invited group Black Lives Matter Toronto activists staged a sit-in which halted the march for thirty minutes.[29] The Black Lives Matter Toronto action functioned as a focal point for the meeting, as community representatives stood to state their diverse responses to this divisive moment for the parade. Documenting the meeting, incorporating these diverse articulations of community at a particular moment, serves to create a record of these multiple articulations of diverse experience as well as facilitating a broadcast space to further develop community responses.

Tara sees facilitating such a broadcast space for community and identity representation as a core value of the show:

> I think giving voice to the voiceless, giving voice to LGBT and other marginalized individuals, I think that's one aspect; gay artists, that's another aspect, gay businesses absolutely . . . and also allies of the LGBT community, getting them involved, getting their aspects of, how do they talk to their gay and lesbian friends, if they have an issue, talk about it—like, come on my show, talk about it. . . . We wanna talk about that sort of stuff. So I don't see it [the show] as having a specific avenue; I think if anything it's more of an open platform that people can put their own stamp on.[30]

However, and crucially, Tara wants to expand the show's listenership and its project of representation and inclusivity:

> What I wanted to use my show for, is to talk to, if you want to call them haters, let's call them haters, what have you, because to be able to understand how they think. And how they feel. And not to *challenge* anyone or say that I want to change you or convert you; I want to understand you. And we may not be able to be accepting of each other or . . . but at least let's talk respectfully to each other, . . . you can come on the show and I want to hear your point of view, *why* you have this point of view and *how* you have this point of view. And I can't wait for something like that. I can't wait for someone to approach me and say 'I want to be on your show, I hate gay people! But I want to talk about it.' Because I think there is worth and there's a lesson in that for us.[31]

In its pursuit of inclusivity, enabled further by its midnight time slot, live broadcasts and additional coverage of key LGBTQ-related events such as the town hall meeting, *Rainbow Country* employs multiple strategies for community building and in particular, enabling what Michael Warner terms 'queer culture building', which he defines as '[n]ot just a safe space for queer sex, but the changed possibilities of identity, intelligibility, publics, culture, and sex that appear when the heterosexual couple is no longer the referent or privileged example of sexual culture' (Warner 2002: 187).

Warner argues the value for queer culture building and the strategies which inform this process lie in the possibilities to 'unsettle the garbled but powerful

Figure 4.3. Pride flag at University of Toronto, June 2016. Photo supplied by the author.

norms' supporting normative sexual privilege (2002: 188). In its multifaceted production of spaces for articulation of LGBTQ experiences and identity formation, *Rainbow Country* enables ongoing capacities for such productive unsettling via recognition of the range of changed possibilities for challenging normative frameworks of representation.

PRODUCING STRUCTURES OF FEELING IN *RAINBOW COUNTRY*

Rainbow Country's project of queer self-representation enables LGBTQ community-building across multiple groups and collectivities, as well as providing a wider forum for discussion and debate. One way in which the show performs this function is by providing broadcast space for the relation of individual narratives which, via the airwaves, can provide a shared point of identification and reference for others—speaking both to people within these communities and those unaffiliated with them—who have had similar experiences. Nick Couldry has described the need for everyone, but particularly those otherwise marginalised in society, to be enabled in providing 'an account of oneself' (Couldry 2010: 7). Such on-air relation of individual narratives has a deep significance for community and coalition building; Collins and Bilge (2016) argue that 'collective political consciousness emerges when people see how their individual life experiences are connected to broader social forces.' This can produce 'an unabashed claiming of a collective identity politics as a vehicle of social commentary' (Collins and Bilge 2016: 120). As noted above, community radio's inherent (structural and material) alternativism produces in turn a more substantial and more fluid on-air space for articulating diverse experiences and for reinforcing a range of identity positions.

The limitations of deriving a sense of belonging and purpose solely from ownership of an individual identity position have been recognised and discussed in arguments challenging earlier iterations of identity politics which primarily emphasised the importance of declaring one's situatedness. The apparent agency and freedom activated by such identity ownership enables a commodification of that same reified identity position: 'It is the market that offers opportunities for the development of the self, including the development of gay [LGBTQ] identity' (Jakobsen 2005: 300). In this formulation, '[c]hange, including liberatory change, when within the discourse of autonomous individualism works for capitalism. Capitalist exploitation allows for individual freedom, but the value of that freedom is defined by regulation, including sexual regulation, that makes the individual open to both exploitation and domination' (Jakobsen 2005: 298). In contrast, I argue that the broadcast space facilitated by community radio can—in places—counter this solipsistic emphasis on the individual, drawing together instead such individual experiences to create, recognise and build on shared commonalities where these are experienced by members of marginalised and minority communities and groups. Potentially building on the articulations thus enabled, alliance politics comprising shared affinities across groups based on shared experiences, interests and practices, can in turn reinforce a sense of a collective freedom, as an expansion of and departure from a 'freedom' enshrined in the individual's performed sense of self—and identity. As Jakobsen argues, 'Freedom need not be defined only by an increase in individualism; freedom could also be constituted through alternative social formations. We can, for example, imagine different forms of freedom, forms based not in individualism or communalism but in [or via] various affinities' (Jakobsen 2005: 298). These alliances and affiliations can be knitted together within an emergent alternative public produced around and in response to an established site of articulated experiences related to shared subjectivities.

As I argue here and throughout the book, in their political but also material emphasis on collective approaches rather than a focus on the individual, community radio programmes—sometimes, in some places, and where encouraged and facilitated by the overarching station structures which facilitate and house them—can constitute a useful shared space where such a diversity of experiences and practices can coalesce in community articulation and in productive in-studio and on-air debate. In its production of an inclusive expressive space, *Rainbow Country*'s emphasis on individual stories about aspects of LGBTQ experiences enables the production of structures of feeling. As conceived of by Raymond Williams and discussed initially in chapter 1, *structures of feeling* provide a way of recognising and thinking through ways in which communities can reinforce a felt or shared and inclusive sense of identity through reinforcing actively experienced manifestations of values as 'lived and felt' (Williams 1977: 132). In considering these values as located in structures of feeling, their resonance and emotional power can be meaningfully retrieved and re-presented back to the community, and deployed in turn to reinforce a sense of community from inside such shared experiences. For many in queer communities, as Michael Warner suggests, queer culture is experienced and understood through 'modes of feeling' that 'can be learned rather than experienced as birthright' (Warner 2002: 198), as LGBTQ everyday practices can include social and political negotiations of discursive and spatial assumptions. Articulation of emotionally meaningful and profound experiences as 'lived and felt' resonate on air and help to build knowledges and understandings of shared experiences, which can then be drawn on in turn to provide a reservoir of particular strategies for negotiating moments of disturbance and discrimination related to perceptions of LGBTQ subjectivities. These in turn can be drawn on emotionally and mobilised to provide and strengthen a sense of LGBTQ community identity and shared everyday cultural practices.

CONCLUSION: ARTICULATING COMMUNITY TRAUMA

This recognition of the importance and centrality of community affinities and shared practices in reinforcing a felt sense of community and belonging, via Jakobsen's emphasis on material social relations (to include both shared interests but also practices) and also Harvey's (2001) assertion that community is made and remade through a series of processes, provides crucial context for Toronto's radio coverage of the Pulse nightclub murders in 2016. The nightclub shootings took place in Orlando, Florida, in June 2016 and were met with shock and horror across Toronto as elsewhere. As heard on both community and public service radio in Toronto, the Pulse shootings represented violence directed at LGBTQ *practices*, and reactions from LGBTQ communities expanded on the real collective need for shared and safe spaces; as Jakobsen argues, 'when people seek relief from the alienation of capitalism they will search for something other than an increase in individual autonomy' (Jakobsen 2005: 298). I suggest that *Rainbow Country*'s structural cultural critique which incorporates recognition of the value of sharing experiences collectively on air, provided a crucial space for articulation of the impact of the Pulse shootings on LGBTQ communities. Tara's description of reactions to the Orlando murders embodies this structural critique across emotional and political registers:

> And then this event happened, in Orlando, and it's messed up in all ways, but . . . to take one person, and one of them, and say, it's his fault, it's missing the mark. Because his fault, it's an underlying layer. But I think ultimately what's happening is, we're so ruled by religion. And we're so ruled by organisations that tell us what's right and what's wrong. Yes, we do have to have laws and all that sort of stuff in place, but I think that when people and when organisations start saying that you as a person, you as a human being, because of who you choose to love, you're less than, or because of the colour of your skin, or because of your sex, your gender, you don't have the right to vote, you're a second class citizen. Those things are wrong.
>
> And as we grow and we wake up as a society and realise women aren't less than, people of colour aren't less than, aren't property, gay people can love and

> marry and be who they are, as we as a society grow up and we realise these things, hopefully instances like what just happened in Orlando and other things, like craziness that happens, because of how we are as a society, I think things will start getting easier. But there needs to be those beacons of light, like individuals that stand up and say 'this is who I am', 'this is what we're about'.[32]

In the recognition of the importance of individual narratives of queer selfhood and the necessity to connect these to felt experiences of subjectivity: of gender, of race, of LGBTQ identity, Tara articulates the value of *Rainbow Country*'s cultural work. These narratives produce and contribute to what Warner describes as 'a context for witnessing intense and personal affect while elaborating a public world of belonging and transformation' (Warner 2002: 199). This context coalesces in 'scenes of association and identity that transform the private lives they mediate' (2002: 57), which are characterised by spatialised practices in identified safe spaces—of which Pulse was assumed to be one. Warner explains the importance of mobile sites for queer culture:

> Queer culture has found it necessary to develop . . . knowledge in mobile sites of drag, youth culture, music, dance, parades, flaunting, and cruising—sites whose mobility makes them possible. . . .
>
> No group is more dependent on this kind of pattern in urban space than queers. If we could not concentrate a publicly accessible culture somewhere, we would always be outnumbered and overwhelmed. (Warner 2002: 202, 204)

Recognition and discussion of the nuances of queer cultural and spatial practices can contribute to LGBTQ community-building and to what Warner terms queer culture world-making. Both *Sex City* and *Rainbow Country* produce spaces for and facilitate in-depth discussion of these nuances in important ways, transcending their more limited articulations in mainstream—including public service—contexts. More broadly, CIUT's provision for community programming across its schedule demonstrates a synthesis between a mobilisation of marginalised subjectivities, which then—ideologically, politically—drive this project of meaningfully alternative articulation.

NOTES

1. My argument here draws on Miranda Joseph's (2002) identification and recognition of this function and mechanism of NGOs, which she develops substantially.
2. Interview with Theresa Lombardi, 15 June 2016.
3. Interview with Lombardi.

4. From CIUT's 'About' webpage, http://www.ciut.fm/about/ (accessed 27 January 2017).

5. Interview with Ken Stowar, 9 June 2016.

6. I introduce and expand on the concept of accented radio (building on Hamid Naficy's theory of accented cultural production, 2001) in my book *Broadcasting Diversity: Migrant Representation in Irish Radio* (2013).

7. Interview with Stowar.

8. Interview with Stowar.

9. Interview with Stowar.

10. Interview with Justin Gunderson, 23 June 2016.

11. Interview with Daryn Caister, 4 July 2016.

12. Interview with Caister.

13. See http://www.bringyaehgame.com/ and http://www.greenmajority.ca/ (accessed 31 July 2018).

14. Interview with Caister.

15. Interview with Caister.

16. Interview with Caister.

17. Interview with Jon Pressick, 23 June 2016.

18. Interview with Pressick.

19. Interview with Pressick; see also http://www.ciut.fm/shows-2/spoken-word/sex-city/ (accessed 31 July 2018).

20. Interview with Pressick.

21. Interview with Mark Tara, 21 June 2016.

22. I use LGBTQ here and throughout this book as this is most commonly in use at time of writing to describe affiliated communities and allies. Tara used LGBT in interview which is replicated here.

23. Interview with Tara.

24. Interview with Tara.

25. Interview with Tara.

26. Interview with Tara.

27. Interview with Tara.

28. Interview with Stowar.

29. 'Black Lives Matter Toronto Stalls Pride Parade', CBC News, 3 July 2016, from http://www.cbc.ca/news/canada/toronto/pride-parade-toronto-1.3662823 (accessed 1 February 2017).

30. Interview with Tara.

31. Interview with Tara.

32. Interview with Tara.

Chapter Five

The Cultural Work of Community Radio

Specificities of Tribal Radio

If the central project of community radio is to enable self-representation through active participation across community-led production, collective feedback and equitable management, this work is particularly valuable for groups in precarious contexts. The three tribal stations I visited at reservations in Arizona combine cultural preservation, educational and first language programming and information provision with an eclectic mix of musical genres across their broadcast schedules.[1] KPYT outside Tucson in southern Arizona, KUYI in northeast Arizona and KWLP outside Kingman in northwest Arizona each broadcast on FM frequencies via a different license type and funding model but utilize common community-building strategies. Tribal stations, defined here as radio stations located in and serving American Indian reservations, or tribal lands, constitute perhaps the largest group of dedicated community radio stations with a common interest in the United States: there are well over one hundred reservation stations across the country.[2] For tribal stations based on tribal lands, the need for community-led programming can be both more pronounced—even urgent—and less resourced than for the community programmes and stations considered so far. An example of the importance of tribal stations for reservation communication is KWLP's and KUYI's incorporation of emergency response functionality, availing of the strong FM signal which renders each station as the most reliable source of information for the Hualapai, Hopi and Tewa tribes they serve.

As with all community stations and shows, engagement with tribal needs and concerns manifests in diverse and specific ways depending on local geographical and material factors. There are material commonalities across tribal radio stations, many of which are on tribal lands experiencing poverty and precarity: the under-resourced infrastructure of many tribal reservations can make transport to and from the station difficult, which then affects volunteer numbers and participation. Despite these limitations, tribal practitioners draw on particularly rich cultural, linguistic and musical traditions for show content, producing diverse programmes with valuable insights into each tribal culture. Technological developments enabling rapid and reliable live online streaming of content have expanded listenership for tribal stations beyond the allocated analogue broadcast signal, meaning that tribal members living and working outside the reservation can still access tribal content on an everyday basis. At KPYT, KUYI and KWLP many shows are regularly broadcast live, particularly the popular daytime shows broadcast during the week, and content incorporates phone-ins, song requests, shout-outs, event promotion and information provision. Crucially for tribal audiences living away from tribal lands, streamed content connects them to tribal culture but also to everyday reservation life. In this, and in the transmission of specific tribal cultural and ceremonial content, tribal stations in many ways epitomise the capacities of the community radio form. In this book I argue that community radio is an essential resource; in this chapter, I suggest the need for such a resource can be particularly acute for tribal communities. KPYT, serving the Pascua Yaqui tribe; KUYI, serving the Hopi and Tewa tribes; and KWLP, serving the Hualapai tribe, each negotiate the interaction between tribal and 'outside' culture in different and diverse ways. Broadcasting tribal content enables listeners both on tribal lands and living outside to connect with and learn from this programming, while at the same time preserving these cultural forms and encouraging everyday use of tribal languages.

As with prior chapters, I situate my discussion and analysis of tribal radio production practices within informing critical frameworks which allow for more thorough recognition and exploration of their particularities. In particular, I draw on First Nations writers and on critical perspectives from development frameworks. Community radio's capacity to contribute to central development aims such as community empowerment is recognised by UNESCO. The possibilities of community radio are summarised by UNESCO as follows:

> What makes a community radio [station] so powerful is its potential to reach out to people with little or no access to information. It is an efficient tool for educating and informing villagers about such critical issues such as health, education, and sustainable development. Community radios are also being used to promote oral traditions.[3]

The fieldwork and research for this chapter has led me for the first time to consider development perspectives and the ways in which these could facilitate understanding of the material and infrastructural contexts within which the three tribal stations operate. At the same time I recognise that

Figure 5.1. KWLP antenna, Hualapai Indian Community, Peach Springs, Arizona. Photo supplied by KWLP.

development approaches themselves are increasingly contested in their imposition of top-down frameworks. I draw here therefore from two approaches which provide a critically reflexive lens: Kate Winskell and Daniel Enger's (2014) unpacking of diverse deployments of narrative and storytelling in development contexts, and Daniel Fisher's (2013) dissection of intimacy and the role of liveness for Australian Indigenous media. What became immediately apparent to me in this final fieldwork for this book is the comprehensive value of community radio's structures and production opportunities for each of the three tribes. Although the communities served by KPYT, KWLP and KUYI negotiate varying degrees of precarity, poverty and isolation, there is notable community participation by my observation in each radio station, reinforcing afresh a recognition of the capacities of community radio for community empowerment, in more pronounced ways than seen in the urban contexts discussed up to now.

CONNECTING WITH TRIBAL CULTURE

KPYT, KUYI and KWLP each produce culturally specific shows for the tribes they serve and represent, combining ceremonial, first language and music programming across their schedules. Through provision of cultural talk and music programming, each station invokes similar strategies in their production of tribal content to reflect the particularities of tribal culture, such as the inclusion and mixing of tribal languages with English on air. The schedule for each station is further shaped by seasonal events and practices for each tribe. Tribal station schedules can access radio content produced by the Native American Radio Network, an organisation producing syndicated radio content by and for Native Americans.[4] KWLP broadcasts *National Native News* from NV1 Mondays through Fridays between 1 and 2 p.m. All three stations broadcast community information via regular public service announcements (PSAs), as part of talk programme content and via Facebook posts. At KPYT, station manager Hector Youtsey confirms the station is the main source of information for the Pascua Yaqui tribe, generally supplanting previous modes of communication on the reservation such as flyers or word of mouth.[5] The modern adobe KPYT station building is located in the Pascua Yaqui reservation outside Tucson in southern Arizona, surrounded by desert hills dotted with cacti in strange twisting shapes which are unique to the area. Just down the road from the station building is the reservation casino. The reservation is delineated by an adobe wall adorned with Pascua Yaqui symbols. KPYT's technical manager Gabriel Otero describes the tribe's reliance on the station for information:

> It's like you really know what's going on in the tribe, whether it's a baseball fundraiser for uniforms to the Healthy Walk/Run to just a birthday shout-out or an anniversary . . . to have a place to do that, is just a beautiful experience for them [community members], and it is appreciated. . . .
>
> In the tribe I believe that most tribal members are very fortunate to have us and they feel that way. Everyone's already for us, everyone already knows they can call us, everyone already relies on us to put their information out.[6]

At KUYI, the specificity of Hopi and Tewa programming for listeners, both on the tribal reservation and living away from it, is important to its practitioners as well as tribal audiences. Tony Dukepoo, the technical manager at KUYI, explains the appeal of their programming for both tribal listeners and new listeners from outside the tribe:

> I think it [KUYI] is a good platform for representing the Hopi tribe, the Hopi Tewa people. It's a radio station geared and moulded to a specific audience. But that specific audience gets larger all the time, so that specific audience meaning our own people; there's a lot of our people out there that are listening now that we're online. So we get all these other individuals that are not even Hopi or anything but are yet still enquiring about songs, content, what's happening out here.[7]

The capacities of online streaming have produced a substantial expansion in listeners for many community stations able to avail of this now established technology. This capacity to share community content more widely is of particular value to those negotiating relocation from a transcultural perspective, for whom community-produced content serves as a tangible reminder of 'home'. Negotiating transcultural experiences is common for many tribal members seeking work outside tribal lands. KUYI producer-presenter Aldric George describes the differences between the 'Anglo' and the 'Bahana' worlds and the need to navigate between the two:

> Being Hopi, I guess it's like, I think what Bruce [Talawyma] was trying to say [in interview] is that you've got one foot—like here's the line, you've got one foot being how you're going to be a Hopi, as a male or a female, and all the responsibilities you'll have as a member of this tribe. And then you've got the other side where it's just the Anglo society, that's our term. In my language it's the Bahana world. And you're trying to navigate that world.[8]

Tony Dukepoo relied on online broadcasts of KUYI for support when he lived outside the reservation, and recognizes the value of such connectivity:

> Especially [for] the ones that are out there in the world, that are travelling all over and listening online. They're able to feel lost out there in the world but then

> they tune into us and they get grounded back to earth again. I guess it enlightens, enlivens them again, wherever they're at. Because I think that's the same type of feeling that I had out there, in the world.[9]

It is the persistent intimacy of liveness in radio content which confers a particular value on what is said and played—whether streamed or broadcast by FM or LP FM antenna—and which can render live streamed content that much more emotionally resonant for listeners living away from their home community or tribal lands. Writing about Indigenous radio in Australia, Fisher unpacks ways to consider how radio's capacity for intimacy further facilitates ongoing processes of community reproduction through ongoing collective strategies described by David Harvey in chapter 1. Thinking through the centrality of intimacy in relation to Indigenous radio production, Fisher suggests that 'intimacy describes internally directed representations of a collective cultural subject'; thus 'intimacy figures as a powerful domain of collective self-constitution' (Fisher 2013: 378–79). In other words, the intimacy possible in live radio enables a particularly resonant expression of community. Live community radio produced by tribal members has the capacity to gather in listeners through familiar content—such as well-known songs, shout-outs to friends and family members or everyday references to reservation daily life. Online streaming expands this reach and strengthens a felt sense of collective intimacy in essential ways for members living 'in the world'.

Tony Dukepoo suggests the tribal music KUYI broadcasts has a particular capacity for Hopi and Tewa listeners to connect with their memories, even as the particularities of certain songs when played on air produce complex responses from listeners: 'The content that we play, how am I trying to say this: the content that we're airing is, I guess it's sensitive to some folks because the songs that are sung, some of them are old, some of them are new, some of them are just memories connected with some of those songs, and they like it'.[10] The range of music broadcast on tribal radio is regionally specific yet considerably diverse in terms of genre. KPYT includes traditional Yaqui music in its daytime programming, which is a version of country music that incorporates the fiddle and was developed in Mexico, where the Pascua Yaqui tribe originated. The shows on the KWLP, KPYT and KUYI schedules play songs ranging from country, reggae, 'oldies', rock and roots. Further genres and styles emerge from specifically tribal musics: for example, 'country rez' is played by reservation bands covering country classics. Sylvia 'DJ Sugarbear' Jackson believes that the selection of music on her KWLP show, alongside her regular use of Hualapai language, connects her listeners both to the familiarity of the music she plays and the memories and traditions around this music. Jackson, who is perhaps better known via her radio name of DJ Sugarbear, plays country and 'rez country' music by Native American country bands; both musical genres are very popular locally, as she says: 'It's pretty good. It's like listening to "country gold" but with our own people, our own Native Americans from Arizona.'[11] Along with Wanda Quasula, who hosts *Native Noon Hour* on KWLP between 12 and 1 p.m. Mondays through Fridays, Jackson also plays traditional and ceremonial Native American music which includes birdsong, drums, powwow singing and other styles used for powwows.

At KUYI, traditional stories are recorded for broadcast according to seasonal guidelines that direct which stories can be told and heard in each month. Station technician Thomas Humeyestewa ensures that seasonally specific stories are ready for broadcast at the appropriate time of year, when they will be most useful, as he explains: 'Sometimes it has to wait a whole year until the time comes around again when I can play it'. The stories are typically retold by elders and recorded by Humeyestewa, either in the elder's home or at the station. A significant part of the structure and form of Hopi storytelling lies in the vocal delivery, which requires that a respondent be present to say 'hui' at regular intervals to acknowledge what is being said. Once recorded, the story undergoes a further hearing from others in the station or perhaps another elder, to ensure there is a consensus on how it has been retold and edited for broadcast. The process is important to ensure community approval and acceptance of the story once broadcast, as Humeyestewa explains:

> We got a round cycle in the year, in the Hopi cycle, when certain things can be played at certain times. And leading up to that, before it gets on air, I get other people's opinions, other elders, to see if it sounds okay to them. That makes me feel more comfortable because when I put something out there that shouldn't be said or done, I get feedback from the community: 'How come you're playing that? You're not supposed to be playing that'; 'you're not supposed to be playing drums during this time because it's quiet time'. Just things like that you've got to think about, when you do recordings like that. That just determines when I get to air it or not.[12]

This set of practices incorporates retelling, recording, listening for feedback, comments and ultimately approval in a collective production of the final version of the story to be aired. This collaborative approach ensures wider approval from tribal members. The activity of providing feedback on the story as it is prepared for broadcast comprises a form of community-building—reinforcing both the story's value and the particularities of the storytelling practice. The consultative feedback process has evolved since its inception and has only gradually become accepted. When KUYI began to record stories for broadcast there was more tribal reluctance over this activity, as station presenter Bruce Talawyma describes, '[T]here was some hesitancy, some thinking behind that: okay, how do we do this, what do we do with that kind of thing. But it was also accepted, because it wasn't being done anymore, in the homes, because it was always in the home [before]'.[13]

The retelling and subsequent consultation process actively involves the contribution of elders as inherent to the practice of retelling for broadcast. The collective production of the stories that are finally aired imbues their public circulation with a further layer of meaning conferred by this mediation process, as the retelling is an active component of the broadcast version. This collective practice enables community-building in substantial ways, as outlined by Winskell and Enger:

> Inherently dialogical, narrative is a primary tool in the continuing social renegotiation of meaning. As something that is literally co-constructed by teller and audience, brought to life by drawing both on the lived experience of the listener or reader and on common cultural models, narrative lays bare the dialectic of human agency and structure in communication. (Winskell and Enger 2014: 193)

The meanings generated by these brought-to-air stories is multilayered. Tribal reproduction of traditional stories brought to the airwaves generates collective agency. At the same time, the history of how stories have been told and retold to children, and the educational purpose of this retelling, sit alongside the meanings embedded in the stories themselves. The Hopi have used the stories to relate practical and ethical lessons for children, as Bruce Talawyma explains:

> At first I don't think it was very well received. . . . Storytelling, the way storytelling is traditional, a long time ago, first of all, it's a way—we didn't look at it that way, but it was kind of a disciplining process for us when we were kids. So what that means was that during this time, this moon, we weren't supposed to be running around. We were supposed to be resting. So that was a way to keep us inside, too. But we didn't think about it that way. So the storytelling was a good way to do that—it was keeping us off the streets, so to speak.
>
> And also . . . it was more or less, it was a combination of things. It was like a history lesson, it was morale-building, it was like, what's the moral of this story, and so a lot of storytellers would say at the end of their story, it was a story about maybe something happened to somebody, so they would say: this is the reason you don't do this! So there was a moral behind the story.[14]

Storytelling is a popular mode of informal education across cultures because it can communicate the central values of a community through an emotionally engaging mode of address, as Winskell and Enger suggest:

> Narrative is closer to non-narrative communication to lived experience in its simultaneous appeal to multiple senses, to reason and emotion, to intellect and imagination, and to fact and value. . . . Narratives are easier to remember and generate emotional responses including empathy and identification which 'allow us to live, experience, and learn vicariously'. (2014: 191)

The stories resonate in layered ways: as narratives, as examples of life lessons for children and as evolving and changing over time, subject to ongoing revision to incorporate new reference points of speaker and audience alike. Once broadcast, the stories resonate at a further layer—as an expression of Hopi culture, bringing together key values in a single story, as Winskell and Enger suggest: 'Narrative brings cohesion to a culture' (2014: 193), reinforcing a sense of community identity. Recording, editing and finally broadcasting these stories on the radio necessarily transforms their telling: as with all radio and media forms, the actions of editing and production for broadcast purposes function to mediate the content going on air.

The on-air sharing of these stories is particularly crucial because the traditional habit of telling children educational stories has been falling out of everyday practice for local Hopi, as Thomas Humeyestewa explains: 'My decision, the reason why I play it [the story], is for teaching. Teaching lessons, history lessons, so people can learn about who we are, where we come from, and like I say, it's not really talked about at home anymore'.[15]

The educational project of broadcasting these stories does not quite replace the family storytelling practice or ritual but perhaps can stand in for it. The broadcasting of stories in modes directed by the tribe also enables listeners outside the reservation to feel connected to Hopi traditions. The retelling of traditional stories through the Hopi language additionally provides a space for learning or relearning the language for tribal members who don't speak Hopi or who speak it less than before. This is particularly true of younger people, as Talawyma reflects,

> [I]t was an educational thing. When you speak Hopi, you understand: everything is clear. Now the different generations, like he said, they don't talk Hopi, they can understand some things. So trying to do a translation based on a recording, it's kind of a challenge. It's not bad or anything but it's challenging. Though you're missing some pieces, but then, the acceptance is there saying, we're not doing it anymore [at home], so that was the idea behind the radio station trying to record it.
>
> So we wanted to keep this going. And sometimes whoever's on air will encourage that and say, do this at home! Because that's where a lot of it belongs. But we're doing it right now so that the kids, the younger generation can keep hearing it, ask questions: how was this done before? You know, that type of thing. Because I tell my kids about that, when we used to all sit down. So it's a

> learning curve for all of us. But it's also a good venue because it's coming out on radio.[16]

KUYI combines Hopi with English in several programmes. The KUYI practitioners I spoke to felt that broadcasting through Hopi is important: to keep the language on air and because—perhaps most crucially—it's the language that confers the clearest meaning to what is being expressed. Talawyma describes the Hopi language's particular capacity for clarity:

> We haven't changed our focus on language. That's one thing we keep harping on, you know, stressing. Teach your kids Hopi. Because that *is* the key to our culture. That's the best way that I can put it, because that really unlocks your knowledge of . . . one word can say so many things, and it goes deep. And it has a lot of emotional compassion, that's what those words mean. When you try and translate it, from Hopi to English, you lose it. There's a big gap. You can't never [*sic*] get it right.
>
> That's my thinking. Because you know, I've tried: I can't do it [*laughs*]. I can use whatever words I know, a dictionary, and all that stuff, I'll never get that true meaning across. I think that's our most important message to elders, parents, grandparents. The kids. You're not going to understand it, especially if I say it in a spiritual setting like the kiva. You're not going to understand what's going on if you don't know Hopi.[17]

The regular use and broadcast of tribal languages on air, for traditional cultural expression such as storytelling but also regularly included as part of broadcast talk content, fulfils a further function of bridging the space between tribal everyday life and what producer-presenter Aldric George calls 'the Bahana world', or the white world outside tribal lands.[18] On air, Hopi is often simply the best vehicle for expression. Talawyma explains negotiating between the nuances of Hopi and using language of 'the other side': 'So when we think about complexity. So you gotta think Hopi. And then you gotta think the other way. So what he's talking about . . . it's pretty simple on the Hopi side. But then when you go to the other side, it's, so, am I wrong or what? [*laughs*] So it's challenging, but it's a good thing'.[19]

As an example of accented radio, this use and foregrounding of Hopi enables listeners to feel connected to the show, both through what is said and the nuances through which content is being expressed. As a function of tribal radio, this use of Hopi is additionally educational for younger tribal members; at the same time, simply speaking Hopi on air reinforces its everyday use by tribal members, which contributes to preservation of the language.

In addition to using Hopi for storytelling, KUYI incorporates Hopi, and occasionally Tewa, language across its talk content, often translating a word or phrase on air. KPYT broadcasts a dedicated show entirely spoken in the Pascua Yaqui language for two hours in the popular lunchtime slot three times a week. KWLP mixes English and Hualapai regularly across show content. For her music show on KWLP, Jackson describes the importance of speaking in Hualapai:

> When I'm talking, I know who my audience is . . . sometimes it's just better to talk in Hualapai than in English because I can just say it better. The people who listen to me, I know who they are; they're older, and probably my age. And then again to the younger people who probably aren't that fluent, just to expose them to the language, they might question somebody, 'what does that mean?' or 'what does that say?' Because I think that if they hear it all the time they might become comfortable and start talking it [Hualapai]. Because no one's forcing you to listen or to be in a classroom, you just hear it every day.[20]

KUYI broadcasts to Navajo (who fall into their geographical signal area) and Tewa language groups and tribal members as well as to Hopi. Consequently, how language and dialects are included and balanced in the schedule is important and must be handled sensitively. Inherent to Hopi storytelling are the nuances of the particular dialect used in the recording, which needs to be recognised and acknowledged: 'That's kind of the way that language is. There's villages that know that same story but they tell it differently; just different ways of telling the story. I've just got to put a note in there: this is a story from Third Mesa, they know this history. Things like that'.[21]

A further particularity of Hopi tribal modes of expression is the way in which everyday information, but also event information, are relayed. This mode of event information provision informs the ways in which events on the Hopi tribal calendar are broadcast by the station. KUYI's presentation of stories, ceremonial and seasonal event information is undertaken with precise attention to what is left out as much as what is included and broadcast. Central to KUYI's delivery and presentation of both stories and community updates are necessary silences and omissions where aspects of a story, or of event information, are related by their absence, with selected particulars alluded to rather than provided in full. For example, for dances scheduled at certain times of the year, the time, place and other details may not be related on air. Instead the dance is referred to obliquely and in passing by presenters:

> You can just briefly mention something, and people that know, that are out there, they'll get the message. For example, now we don't announce dances. We just make comments like, this is time for the village, things are happening, so you know what they are, kind of thing. Since we're talking Hopi, we can say

the words, but we're not going to go into any more detail than that. So they pick it up. They get that.[22]

The provision for produced silences is characteristic in some ways of the Hopi approach to talking on air. At KWLP, they encourage Hualapai community members to call in to a given show with an anonymous shout-out, for those who do not want to hear their name or voice on air but who still want to send out a message. Several people at KWLP spoke to me of the shyness of some Hualapai about being heard on air and described ways in which this shyness was provided for in the daily shows, in strategies such as the anonymous shout-out. These culturally specific communicative nuances are as important as the content they inhabit, as they too are shaped by tribally specific traditions of spoken delivery. Thus, the very mode of speaking is of the tribe—making this content both accessible to tribal members but also respecting tribal delivery modes.

This embedded situatedness is characteristic of and inherent to the functions of accented radio as discussed earlier in this book, and can additionally be read as a decolonising mechanism enabling tribal self-representation. Jim Cheney (2005), drawing substantially on American Indian formulations, suggests that tribal situatedness is inherently relational rather than hierarchical. In this paradigmatic understanding, the music, rituals and stories which comprise what he terms 'ceremonial worlds' are important in their own right and also centrally linked to tribal identity. A collective and holistic American Indian situatedness is in clear contrast with the dominant and overarching US paradigm of neoliberal capitalism with its central tropes of individual achievement and systemic rationality. Thus, articulations of tribal identity via particular tribal cultural expressions can simultaneously reinforce tribal situatedness and serve as a counterpoint to the hierarchical logics of neoliberal discursive framings of 'Native American culture' from the outside. These articulations can therefore challenge reductive representations of tribal culture and peoples which circulate in US mainstream media. Taiaiake Alfred (2005) explains the importance of working with a 'decolonising framework' in what he terms an 'indigenous resurgence'. For Alfred, a decolonising framework is built on retrieval of all parts of American Indian culture, including reaffirming tribal relationships with the land, tribal culture and languages, and teaching these to the youngest tribal generation to ensure continuity. Thus broadcasted tribal content reaffirms tribal identity and experience at the everyday level for regular listeners—but also at a sedimentary level, where and from which tribal traditions can be revitalised, through such regular normalisation which in turn re-creates and reinforces a collective familiarity.

Figure 5.2. KWLP outside broadcast, Hualapai Indian Community, Peach Springs, Arizona. Photo supplied by KWLP.

MOBILISING REMITS AND EXPANDING EDUCATIONAL PROGRAMMING

The three stations operate under different license categories and funding structures. KPYT receives funding from the Pascua Yaqui casino on the reservation, an impressive building half a mile from the station. This relative financial security includes the provision of standalone studio and station space and confers further stability in relation to the everyday running of and resources for the station. KUYI unusually broadcasts via two frequencies: an FM frequency and, since 2017, an additional low-power (LP) frequency, which the station applied for to ensure coverage of areas which were not receiving broadcasts from the FM signal. This is necessary given the substantial scale of the Hopi reservation the station broadcasts from and to, as well as adjacent Navajo and Tewa tribal communities. At the time of my visit, KUYI was temporarily housed in a building incorporating studio and meeting spaces and another studio for production. The Hopi reservation in northeast Arizona incorporates First, Second and Third Mesa lands, which each carries its own traditions and also borders on Navajo lands. The nonprofit category of the station's license means that KUYI necessarily incorporates ongoing fundraising activities in its yearly calendar such as calls for donations and fundraising events, which are published on the station's website and Facebook page. In contrast to KUYI's struggle to acquire comprehensive broadcast coverage via their two frequencies, KWLP is relatively fortunate as the station secured a Tribal Priority license for their tribal station in 2013. Tribal Priority licenses were introduced in 2010 by the FCC: 'The Tribal Priority is intended not only to provide radio service tailored to specific Tribal needs and cultures, but to increase ownership of such radio stations by Tribes and Tribally owned entities. In doing so, the Tribal Priority also fosters localism and diversity of ownership'.[23] Working with tribal elders, station manager Terri Hutchens oversaw the application and acquisition of the KWLP's Tribal Priority license, a process in which 'everything was subject to approval from the tribe'. KWLP's Tribal Priority license is one of only two awarded by the FCC in 2013. She describes KWLP's license structure and remit as unique: 'Ours is really a combination of a traditional . . . small market commercial station and a community LP station. We've kind of merged it in how we do the governance and the management and operations and also the programming and content. It's a very unique merger'.[24]

KUYI broadcasts as a nonprofit station under the heading of 'educational radio'[25] while KWLP began broadcasting online first, producing educational content for young people. Both stations began with a remit for educational programming, an emphasis which continues to inform their programming content and schedule composition. Alongside this educational provision is recognition by station practitioners of ways in which education—both formal and informal, as in confidence-building—can benefit the tribe and its young people in particular. Educational content and training are thus mobilised as a form of empowerment for tribal teenagers; KWLP has retained several young volunteers who have been involved with the radio station since its inception and still present shows years later. Hualapai teenagers are encouraged to produce show content reflectively, creating shows which are meaningful to them. Often these resonate with the community, as with the *Rock Out Hour*, presented and produced by 'DJ Pillow'—Willow Majenty, a young woman who seeks to honour her dad through playing his favourite rock songs.[26] Her show is broadcast for an hour each Tuesday afternoon at 2 p.m. 'DJ Getta' (full name Georgetta Russell) has been with KWLP since she was a teen volunteer at both KWLP and with local youth support groups. DJ Getta presents on three programmes reflecting a wide range of music: *The Morning Blend* (on the popular 10 a.m. to noon daily time slot), and *Native Star Nation* and *In the Studio* on Saturdays.

KUYI produces an educational, teen-focused programme, *The Hopi Teen Show*, which involves local Hopi and Tewa teenagers in producing and presenting programme content. Participating students study at the local high school which works with KUYI to provide opportunities for interested teenagers to get on air and to learn some of the Hopi language. Through their production work with the *Hopi Teen Show*, the students gain confidence and also learn the importance of articulating themselves and their needs and concerns in effective ways. Tony Dukepoo explains how working on the show empowers the students to 'help them be more confident, stronger speakers, because in the world out there you've got to be able to speak your mind, you've got to be able to say something, and if you're not communicating, your enunciation needs to be there or people won't take you seriously'.[27] In identifying the wider benefits for confidence-building through learning radio production and presenting skills, Dukepoo connects the students' acquired experience with the use value of being a confident speaker (and person) in 'the world out there', recognising how valuable this confidence, and display of confidence, can be for these future adults on and outside the reservation. Considered alongside tribal language programming, cultural history and the everyday life lessons for children embedded in Hopi storytelling, these directed programmes serve as a valuable learning and training ground for teens.

For the Hualapai, KWLP functions as a key resource for tribal children and teens in an ongoing working relationship actively promoted in the community. The radio station functions as a hub for the Hualapai community, as explained by Peach Springs School Superintendent Jaime Cole,

> I think the radio is just a neutral ground for everyone that everyone enjoys . . . Everyone in the community, their interest is on kids, so why not use the radio to connect that with parents just to show that our best interest is promoting what's good for kids.
>
> In reality, because the community is so small, everyone has ownership and partnership with the development of kids, because that's going to be the future of the whole community.[28]

Cole's acknowledgement of the importance and centrality of radio for the community echoes the recognition of the valuable ways in which programme content, tribal participation and community reliance coalesce and strengthen connections between the station and community members. Identification of the value of these connections recurs in the insights from the tribal radio practitioners I spoke to—we would often begin talking about a particular topic such as storytelling (for example) or dedicated youth programming—and the practitioner would expand discussion to observe how educational content (for example) enables a sense of connection to the wider community.

At KWLP, station manager Terri Hutchens and station producer Daniel Ondrejka work consistently with station volunteers to fine-tune and maintain broadcast standards of individual presentation style and delivery. Feedback is provided to volunteers in a peer review process at regular mandatory station team meetings. As part of the feedback process, Dan plays a clip from a recent show by each volunteer. The clip is then reviewed by the group according to a set of rigorous criteria used by commercial stations to assess individual broadcasters. The comprehensive presentation criteria include: good degree of preparedness; ensuring a lack of audible breathing; good levels of pace/tempo/articulation/enunciation; lack of 'clutch' words or phrases, stuttering or dead air; good use of station IDs; vocal volume and naturalness of DJ sound.[29] The rigorous criteria enable volunteers to acquire solid presenting skills while simultaneously further legitimating the work of on-air presentation. Should the volunteer find new work in broadcasting, the rigorous training of these feedback sessions prepares them for future radio roles and for similarly rigorous assessment contexts elsewhere. The peer feedback provided at these sessions, by my observation, was both generous and rigorous. Participating volunteers are familiar with the process and thus attuned to the nuances of the feedback criteria and present their comments to their radio colleagues in positive ways—often while laughing about their own performances.

At all three stations, storytelling and other specifically tribal content is produced and broadcast alongside dedicated pedagogical content. The work of reproducing culture—through pedagogy, through collective storytelling, through shared modes of tribally specific delivery and address contributes in turn to community-building; as Winskell and Enger argue: '[c]ulture-making is, by extension, a form of active citizenship' (2014: 194). These practices constitute meaningful and active forms of community participation, a more realised set of ongoing practices as distinct from ad hoc participation activities selectively directed by top-down media structures. Considering the function(s) of pedagogic practice in particular, Winskell and Enger insist that pedagogy 'also serves a political function, undermining the role of authority within the culture. A participatory approach in education, then, prepares learners to be culture-builders, vectors of social change and, by implication, active citizens in a participatory democracy' (2014: 194). At the same time, the particularities of community modes of radio production enable reflexive cultural production specific to individual tribal expression, typified by 'a high degree of reflexivity and an attention to the ways its address presumes a particular cultural and political subject', as Fisher conceives of Indigenous radio (Fisher 2013: 379). Such reflexivity informs and shapes the production of tribally specific content at all three stations from educational programming to collectively produced storytelling.

Figure 5.3. KWLP billboard, outside Hualapai Indian Community, Peach Springs, Arizona. Photo supplied by KWLP.

'VOLUNTEERS ARE THE COMMUNITY'

KPYT, KUYI and KWLP, like most community-run stations, rely primarily on unpaid volunteers to present and produce individual shows. At KUYI, Aldric George presents a weekend music show but has other crucial tribal, farming and work responsibilities which he balances alongside his broadcasting commitment. George acknowledges KUYI's flexibility around his scheduling conflicts and recognises the importance of this for his continuing work for the station:

> [The station manager's] been really great in understanding that I have to balance my time in the Bahana world and in our own, in our culture, the stuff I've got going on for myself. He's been really understanding in allowing me to come back as if I'd never left . . .
>
> The radio station is pretty accepting, I guess accommodating in how we do things, as long as we do it right and the stuff we're putting out there is just what the public wants to hear. I think there's a lot of people who want to hear us speaking in our native language on the radio.[30]

Tony Dukepoo insists that the volunteers are the core of the station and of KUYI's programmes:

> I thought community radio was like commercial radio, where they could get all the money they need through commercials—but after being educated more on community radio that's when I came to conclude that volunteers are pretty much the community and the public is pretty much the strength behind community radio. And from there it made me feel like, it's only going to be as good as we can make it, how hard we can stand behind it and be the backbone for it—because that's technically what it boiled down to: all this assistance with DJs and such, is pretty much the backbone of community radio. All the volunteer DJs, all the volunteers, are pretty much the backbone.
>
> That got me thinking about how we are as people, as Hopi, as Native Americans, as indigenous people, none of us really . . . when we work for something as a community we don't expect things in return. We do it out of our own hearts, out of our own kindness, and in turn that makes anything stronger if we go into a job like that with that kind of mindset.[31]

Both George and Dukepoo expand their discussion of their work to incorporate recognition of the collective work of the station. In his observation, Dukepoo sees the collective work of KUYI as representative of Hopi culture and values. At KPYT, station practices are recognised as subject to the calendar: in particular, every spring, volunteering practices are amended for the duration of Cuaresma. During Cuaresma (familiar as Lent in English) food and drink restrictions are observed at KPYT alongside recognition of changes to work practices. For the Pascua Yaqui, the Cuaresma period also includes the Deer Dancer fiesta and procession, an established tribal tradition which is held on the reservation each year.[32]

All the radio practitioners I spoke to in researching this book spoke of their love of working in radio, perhaps especially those practitioners who were volunteers. As seen throughout this book, individual motivations for volunteer radio work vary. However, across all responses, the investment of time and energy put into radio production and/or presenting was a voluntary choice driven by passion for the work. In diverse ways, the community radio practitioners I spoke to are 'enabled in their projects, projects that are often important to the very survival of the people served by the organisation', as Miranda Joseph describes volunteers at nonprofit organisations (Joseph 2002: 74). Joseph suggests that nonprofit workers who volunteer, far from being passive subjects at the service of the overarching nonprofit where they work, are instead actively involved in their volunteer work, believing (but often without naïveté) in the wider projects they are working to support. Community radio stations can also provide a supportive working environment for volunteer practitioners, as with Tony Dukepoo's experience working at KUYI:

> With community radio, if you fall at least you have somebody there with you to help get you back up and get going again. And at the same time maybe there's someone there who sees both you two struggling and helps pitch in as well. Whereas commercial, they're not getting paid for it [helping others], they're not going to help you.[33]

For KUYI and KWLP, volunteers can be difficult to recruit and retain. Difficulties in attracting and then retaining volunteers may include logistical factors such as a lack of available transport to and from the tribal station, alongside other factors such as some community members' concern that they might be perceived as competing with other tribal members, and/or individual shyness. Additionally, as discussed above, many tribal members do not necessarily speak the tribal language, meaning that volunteers for tribal language shows can be especially difficult to recruit.

The 'Narrative Statement' in KUYI's 2013 Station Report, drawn up with the Hopi Foundation which supports the station, describes the organisation's perception of the main challenges for volunteer recruitment:

> The Hopi Foundation seeks applicants with experience and skills to meet the minimum requirements of its entry-level positions. Another desired skill is for candidates to speak and/or understand the Hopi language due to KUYIs [*sic*] strong consideration for local programming and goal to broadcast in the Hopi,

Tewa and English languages. While candidates who speak and/or understand the Hopi language are highly considered, they are often those that have limited skills and experience for professional level positions.[34]

The second challenge for hiring employees is the limited access to transportation to and from work. A significant portion of KUYI volunteers and local residents do not own personal transportation. The Hopi Reservation spans 75 miles across twelve major communities with only one minor public transportation system that serves the entire reservation on a limited basis.

The statement incorporates recognition of approaches to overcome these challenges, however, albeit alongside an acknowledgement that an ongoing lack of funding means that more substantial training and outreach initiatives remain beyond the scope of what is possible.

While the Hopi Foundation continues to seek ways to overcome employee recruitment challenges and improve its outreach strategies, including the local high schools [*sic*] radio class, a long-term training program for future professionals in public radio, recruitment for new KUYI staff positions in 2011 has not been attempted due to limited funds available.[35]

Tony Dukepoo and Thomas Humeyestewa at KUYI describe how they replace criticism received from tribal members, calling on people who comment on programme content to come along and volunteer themselves. Dukepoo observes that in relation to criticism from listeners, 'you kind of hear it all out there in the community', and says those who work at KUYI use community feedback as a chance to invite community members to contribute, or simply to visit the station:

[We say:] Come help us out, make it better—with your assistance we can make it better, find those flaws and overcome them. Sometimes they come through, they come by, fill out an app [volunteer application], show up for a couple of days. Sometimes they come by and just bring food or bring goodies, just to show their appreciation, because they've seen that we've bettered ourselves, bettered the station in one way or another. Somehow it all ties together.

They receive complimentary feedback too, as Dukepoo points out: 'with our listening bases, we always get support from them. They just call in and let us know we're doing a good job and that's always good to hear, especially from our listeners'.[36] Community radio's capacity for providing a space in which community feedback can be spoken and responded to is what makes it such a useful and valuable community resource: it produces a working form of outreach, as defined by First Nations activist Harsha Walia,

> Whatever form of outreach we choose, the key is that it should be frequent and with the intent of engaging with people. . . . We need to stop and talk to people, ask them about their opinions while sharing our own, and invite them into spaces where they can expect not only to be talked at but can also provide their own thoughts. (Walia 2013: 181)

CONCLUSION: DEVELOPING TRIBALLY SPECIFIC TOOLS

I began this chapter by proposing that tribal radio can be said to epitomise the importance of community radio's community-building work—in its multiple, diverse forms and formats—and in fact expands on its core tenets of accessibility, accountability, inclusivity and reciprocation in under-resourced reservation contexts where the work of community radio is more urgent. KPYT, KUYI and KWLP each utilise the capacities of community radio to create tribally specific forms of programming via community-led production practice. The diverse licensing categories of the three stations further demonstrate the specificity of each station's negotiation of legislation, frequency provision and ongoing material and resource support. In 1996, the following challenges for Native American radio were identified: 'Lack of money; inadequate staffing and training; lack of Native-language skills; poor facilities and equipment; lack of Native programming; competition' (Keith 1995: 113–23).[37] This categorisation of the multiple problems faced by individual stations certainly applies to many community stations beyond tribal broadcasting: negotiating these problems is a common experience for most community radio stations and practitioners.

Given that community radio is radio produced on the community's own terms and shaped by local materialities including available infrastructure, resources, volunteers—this necessarily collective, grassroots approach is that much more necessary where definitions and understanding of tribal communities remain contested, undermined and denied, as they are and have been for Native American tribes. Each of the three tribal stations I visited produces content and encourages production practices which reinforce and reaffirm Hualapai, Hopi, Tewa and Pascua Yaqui culture and values—not simply through provision of educational and language programming but through production practices embodying tribal modes of expression, respect and silence. In discussing their production and station work, tribal practitioners reference the collective culture of the tribe and talk about their work in terms of how it serves tribal culture, traditions and values. Linda Tuhiwai Smith argues for the importance for 'a critique of how we, as the Other, have been represented or excluded from various accounts' for Indigenous self-determination and insists that 'Indigenous peoples want to tell own our stories, write our own versions, in our own ways, for our own purposes' (Tuhiwai Smith 1999: 28). In diverse ways, KPYT, KWLP and KUYI facilitate space for such a critique by producing tribal content on their own terms.

Central to how tribes perceive themselves and their tribal identity is their relationship with the land (Cheney 2005; Alfred 2015), which is fundamentally different to hegemonic definitions of tribal land in the US. For tribal members it is important that stations are located on tribal land, both geographically and symbolically. For all analogue community stations, location is crucial to fulfil a remit of local proximity and community-facing service, including providing a public space for community members to drop in. Cheney, working with American Indian activists and academics, suggests that in American Indian culture, 'nature's integrity becomes the model of individual and social integrity' (Cheney 2005: 121). Expanding on this, he suggests that 'Indian knowledge' and experience of living in the world is based on a respect for it, 'requiring attentive listening to, and reciprocal communication with, the earth and is woven together within ceremonial worlds designed to accommodate human culture with (and *as*) a wild world' (Cheney 2005: 120, emphasis in the original).

Through this perspective the inclusion and broadcast presentation of aspects of ceremonial culture such as powwow music and storytelling can be considered cultural preservation, but also a reinforcement of principles of tribal life and philosophy. Gathered through tribally specific production practices—such as Humeyestewa's recording of stories told, and approved, by elders in accordance with traditions of both telling and hearing—these three tribal stations reaffirm and reinforce tribal community identity through both formal and informal broadcast content. For Cheney, referencing Native American accounts, 'ceremonial worlds' situate 'communication and reciprocity with natural environments, rather than the desire to dominate those environments, at the very heart of the production of knowledge' (Cheney 2005: 124). Invocations of reciprocity and collective communication reverberate in tribal practitioners' descriptions of their work, reinforcing the importance and centrality of these values for the tribe.

Finally, in this book I repeatedly argue for community radio's political use value: not exclusively or primarily as a sounding board but as a communal location—comprising both physical studio space and meeting spaces, but also and simply a shared broadcasting 'space' for community self-representation. Marginalised communities often struggle on multiple material and political fronts as a result of a substantially marginalised, even disenfranchised subjectivity. At its best, community radio provides tools for negotiating this struggle in the form of accessible and useful content for tribal members. These are 'strategies that expand prefigurative spaces', as Walia argues, where 'we reduce our reliance on the state and strengthen our infrastructures of resistance by creating media, collective kitchens, infoshops, and so on, are truly revolutionary when they connect with as opposed to withdraw from people' (Walia 2013: 181).

In providing the means to articulate and discursively participate in this struggle, community radio and tribal radio's capacity for such ongoing outreach strategies demonstrates its value. In my observation, these three tribal stations reinforce an empowering sense of 'community' in their tribally specific production practices, instead of adhering to top-down modes of production lacking in both familiarity and tribal recognition.

NOTES

1. In this book, my uses of the terms 'American Indian', 'Native American' and 'reservation' have been informed by the practitioners I interviewed from the three radio stations discussed here. All three terms are used by some and contested by others. Therefore, I have used American Indian and Native American interchangeably in recognition that they are in regular use by many tribal members. 'Reservation' is used here to describe the physical land allocated to the tribal communities under discussion, but I also use 'tribal lands' as a possible alternative. Wherever possible I have referenced either the specific tribe being discussed or used the word 'tribal' as this is what I heard most and seems the most specific description of the communities.

2. The Native American Radio Network, also operating as Native Voice One, comprises affiliate stations serving tribal communities across the United States; see http://www.nv1.org/stations-affiliates/ for a list of affiliate stations (accessed 24 July 2018).

3. Description from UNESCO, https://en.unesco.org/news/community-radios-promote-local-development-across-africa (accessed 10 August 2018).

4. The Native American Radio Network 'distributes work from Native and non-Native producers whose programming educates, advocates, and celebrates Indigenous life and value' syndicated across both reservation stations and some National Public Radio affiliate stations. Description from the About section, http://www.nv1.org/about/ (accessed 10 August 2018).

5. Programming details from KWLP schedule from http://www.epchradio.com/index.php/station-info/daily-programing (accessed 15 May 2017).

6. Interviews with Hector Youtsey and Gabriel Otero, KPYT, 7 April 2017.

7. Interview with Tony Dukepoo, KUYI, 8 May 2017.

8. Interview with Aldric George, KUYI, 18 April 2017.

9. Interview with Dukepoo, KUYI, 8 May 2017.

10. Dukepoo, KUYI, 8 May 2017.

11. Interview with Sylvia Jackson, KWLP, 13 April 2017.

12. Roundtable interview at KUYI, 18 April 2017: insights into recording and production of story broadcast from Thomas Humeyestewa.

13. B. Talawyma, KUYI roundtable interview, 18 April 2017.
14. Talawyma, KUYI roundtable.
15. T. Humeyestewa, KUYI roundtable interview, 18 April 2017.
16. Talawyma, KUYI roundtable.
17. Humeyestewa, KUYI roundtable.
18. Interview with Aldric George, KUYI, 18 April 2017.
19. Talawyma, KUYI roundtable.
20. S. Jackson, 13 April 2017.
21. Humeyestewa, KUYI roundtable.
22. Talawyma, KUYI roundtable.
23. See https://www.fcc.gov/news-events/blog/2013/03/01/tribal-radio-priority (accessed 10 August 2018).
24. Interview with Terri Hutchins, KWLP, 13 April 2017.
25. License and provision details from the 2013 EEO [Equal Employment Opportunity] Program Report for KUYI, see note 34.
26. As told to me during a station meeting at KWLP, 13 April 2017.
27. T. Dukepoo, 8 May 2017.
28. Interview with Jaime Cole, 23 April 2017. Cole also produces and presents the *Drive at Five* show for KWLP broadcasting on Fridays at 5 p.m. and previously hosted the *Live at Five Reggae* show on Thursday evenings.
29. Categories listed as they appear on the KWLP's 'DJ Voice Check/Training Checklist' feedback document, as circulated in a staff meeting I attended in Peach Springs, Arizona, on 13 April 2017.
30. A. George, 18 April 2017.
31. Dukepoo, 8 May 2017.
32. Interview with Gabriel Otero, 7 April 2017.
33. Dukepoo, 8 May 2017.
34. From the 'Narrative Statement' in the 2013 EEO Program Report for KUYI, accessed from http://www.kuyi.net/kuyi-announcements on 9 May 2017.
35. From the 'Narrative Statement' in the 2013 EEO Program Report for KUYI, as accessed above.
36. Dukepoo, 8 May 2017.
37. Each of these is a section heading in the chapter 'Traversing the Gorge: Challenges for Indian Broadcasters', in Keith's comprehensive 1995 study *Signals in the Air: Native Broadcasting in America*. Westport, CT: Praeger.

Conclusion

Practices, Publics, Policy

I began this book by explaining the significance of the inclusive project of community radio: for enabling and reinforcing a necessary alternative space for marginalised community expression, for empowering community practitioners, and for facilitating collective and inclusive practices. At this point I want to pull together my explorations of modes of community articulation and self-representation to revisit how we can critically consider community radio's wider societal contribution in order to identify how the inclusive cultural work of community radio can be mobilised to facilitate and encourage codes of good practice. In this conclusion, therefore, I revisit commonalities of community radio practices as a galvanising point of departure. I firstly re-examine the labour of cultural production undertaken by volunteer practitioners out of their passion for the work—their desire to represent their communities and share knowledge—with a view to consider this cultural labour in ways which recognise its value and importance. Following this, I explore the diverse ways in which community radio facilitates the creation of alternative publics, which themselves enable meaningful participation through shared recognition and negotiation of specific community experiences. Finally, I draw these two exploratory and directive arguments together to unite in a third, in which I discuss recent community initiatives involving sharing good practice in US tribal radio contexts, and developing alternative policy processes for Indigenous broadcasting in Canada,[1] to map ways in which community radio practices can be further legitimated and supported in legislative and policy formations. In this summative examination I consider the cultural labour of community radio production as a multifaceted form of meaningful *participation*; theorise how alternative publics produced in community radio facilitate in-depth, multilayered and nuanced community *representation*. I then suggest ways in which reflective exchanges of good practice and collective approaches to policy formation can be mobilised to further enable both community participation and self-representation in materially and politically supportive and empowering ways.

The work of culture has ever been to comment on, reflect, reproduce and/or challenge the wider world around us. What I am interested in with this project is what Mark Banks terms as 'the capacity of culture to help manifest and enact different kinds of social action' (Banks 2017: 2). To enable a holistic and thorough examination of this capacity, the value(s) of cultural work need(s) to be identified so as to understand how these can more effectively be deployed. As Banks puts it, this

> involves consideration of cultural work as a source of (not just) an economic value, but also a social value, as well as an aesthetic one, underwritten by different kinds of political sensibilities. Doing justice to cultural work therefore means respecting the 'internal' goods and qualities of work as a practice—but without discounting the 'external' structures and pressures that tend to make such work somewhat less than appealing, and often deeply unfair and unjust. (Banks 2017: 2)

Insofar as all the community practitioners I talked with enjoyed their work and felt fulfilled by it, I reproduce the above in full in order to include recognition of the structural unfairness of most community radio production work given that it is unpaid. Station management staff, in contrast, are salaried. Banks's summary therefore combines recognition of ways in which community radio production has purchase in the economic (because providing free labour), the social (because enabling community representation) and the aesthetic spheres (because producing diverse expressive forms and styles). In identifying and exploring the cultural work of community radio programmes and stations, I have actively sought to understand how far this cultural work goes to counter gaps and in community recognition imbalances in which the work of many, even most community practitioners in a context remains undervalued. My examination of the wider social and political contributions of community radio develops in relation to Nancy Fraser's (2008) typology of recognition/representation/redistribition in order to understand and ground analysis of the depth and degree of the cultural goods produced by these programmes for both their producers and for the community(ies) represented. While community radio production practices are flourishing in many places, the structural institution of community radio continues to face substantial struggles, especially with resources and recognition. Crucially, the central component of community radio production remains a dedicated form of labour which is almost everywhere unpaid.

CRITICALLY SITUATING RADIO PRACTICES

Radio practitioners in community contexts almost always produce their shows on a volunteer basis. As demonstrated in my prior chapters, the cultural work of radio production, particularly of a community-facing show, incorporates (but, as the phrase goes, is not limited to) community outreach (equitable representation across community groups; information gathering and provision; organizing free events), interviewee access and interview scheduling, programme structuring and planning, music choice and playlist planning, ongoing time commitment (over months or years) to a broadcast schedule, time management for studio access and topic selection and research. Once in studio, the producer-presenter will often present live, conduct interviews, cue the music, play station IDs and public service announcements, respond to feedback on Facebook and/or Twitter, and take phone calls (if station policy and usual practice permits this). Practitioners decide on topics, guests and items, drawing on in-depth community knowledge in their decision-making. They work within their self-determined programme remit, which is in turn inflected by a politicised sensitivity as to how best to represent and balance community issues, and their production practice is further informed and structured by station practices and guidelines.

This is complex, multifaceted and specialist work. It is valuable labour for both practitioners and the community(ies) represented, and, as I argue, for society as a whole. Community radio production labour can be considered as *contributive* to society at both individual and collective levels. Paul Gomberg (2007) presents and explores contributive work as a 'constellation of goods' (2007: 66) with benefits to the individual and society. Complex work which functions to reinforce a community sense of belonging contributes to both the community represented and the culture more broadly. This work simultaneously enables the practitioner to make an individual contribution produced through and demonstrating their own development of complex abilities and skills, resulting in a programme that is valuable to both the individual and society. Within this intersection of overlapping goods, such individual development enables the practitioner to access and avail of additional and wider opportunities to make further contributions.

For almost all community radio practitioners, this complex labour is carried out without pay. There are significant adjacent rewards in the form of acquisition of cultural capital (through sharing specialist knowledge of music or arts on air), social status (rewarded via insider status on a cultural or music scene and/or acting as a community representative), and/or opportunities to develop and sharpen the array of abilities and skills needed for live radio production. The work of producing a show carries a set of corresponding values for the practitioner which range from the cultural capital of 'working in radio' to the acquisition and sharing of specialist knowledges, to the political capital of representing segments of a given community. The question remains, however, to what extent, if at all, do such acquired values compensate for doing such specialist, time-consuming and time-specific work without pay? Many community radio practitioners produce (and research and present) their shows live in a regular time slot for years. In addition to the substantial labour involved, this commitment also enacts travel costs and requires that practitioners make a dedicated space in their work and life schedules for production and broadcast of their shows.

What is the value of this work and how is its value determined, considered and legitimated for committed volunteers? For the practitioners I talked with, the value of their work is realised in their own passion for and dedication to their respective shows. The shows would not be broadcast in their well-researched and thoughtfully planned form without the sustained commitment of their producer-presenters. The representative and political value—awareness raising and information provision—of community-led programmes is substantial; given this, can the unpaid nature of this work be justified? The community sphere has an established history of reliance on volunteer labour and community radio in particular profoundly relies on it. What contributes to this is recognition of the validity and value of the community radio project itself: characterised variously but consistently as inclusive, community-driven and therefore representative of community culture, needs and concerns, it provides a social good. Writing about expectations around volunteer labour in the cultural industries, Banks suggests that in cultural forms 'where capital investment or subsidy is lacking, and making money not a principal [*sic*] criterion or expectation . . . attitudes tend to be more pragmatic and sanguine about (if not always tolerant of) the need to work for free' (Banks 2017: 136). It is perhaps community radio's pragmatism, embedded in the ongoing requirement to ensure regularly scheduled programming in accordance with a carefully balanced schedule of volunteer practitioners, which enables institutional and individual acceptance of the conditions of volunteerism that typify almost all community radio programme production. Within these structural conditions, it is usually the case that all producer-presenters at a given station are working on a volunteer basis—or in cases where producer-presenters are being paid, they are being paid for other work at the station, such as producing station promos, sound editing work and providing production training. To separate out and identify how components of community radio production map on to complex values, I have assembled a working typology.

- Attainment of cultural capital through sharing acquired specialist knowledges (e.g., activist experience, arts and music knowledge, insider 'cool' knowledge);

- Attainment of cultural capital through the learned capabilities (production and technical experiences gained) of producing a radio show;
- Attainment of community respect through representing that community;
- Attainment of personal satisfaction/fun and self-worth derived from acquisition of skills;
- Recognition and fulfilment of time-specific production responsibilities;
- Acquisition of transferable production skills, making practitioners more competitive for future paid radio positions.

This typology of community radio labour is extrapolated from circulating hierarchies of cultural labour alongside practitioner accounts of the cultural/experiential/political value of their work. Unpacking these diverse areas of enjoyment, fulfilment and satisfaction enables a more nuanced consideration of the diverse value practitioners find in their volunteer radio work. As an illustrative example, Fabiola Charles, Rhonel Cinous and Rebecca Laratte from Miami's *The Morning Drive* embody several of the above categories in their production practice and they are clear about how they see their experience improving access to future opportunities in paid radio work. While they enjoy their work on *The Morning Drive* at creative, expressive and representative registers, they also recognise the value of such acquired skills and experience for their future careers in broadcasting. Community radio production's value can include 'the enjoyment of complexity, of contributing to others, and of being respected for this contribution'; together these embody a 'constellation of goods—contributing developed abilities to the good of others and earning esteem for those contributions' (Gomberg 2007: 66). Building on this, Gomberg developes what he terms a 'constellation of opportunities' linking together activities which are characterised by and circularly produce complexity, contribution and esteem.

A further and crucial dimension to this work is practitioners' individual commitment to the ongoing workload of production as it generates an experiential opportunity provided by acquiring skills and expertise in radio production alongside the development of additional experience in carrying out community outreach work. Media and radio production experience are increasingly in demand as the cultural capital of 'working in the media' continues to increase. As with other areas of cultural work considered as desirable and fulfilling (see Joseph 2002, Banks 2017), opportunities to acquire sought-after production work remain limited, especially as they remain available to young people from privileged environments far more than people from more marginalised groups and/or environments. Therefore, a considerable incentive for new radio practitioners to do volunteer work in radio production is realised in the chance to gain this experience and acquire a new array of skills and experience. Rosalind Gill and Andy Pratt (2008) call for recognition of 'new forms and practice of politics that are mobilised around the precariousness that is said to be a defining feature of contemporary life'. Through a critical examination of this informing condition of precarity, they advocate a critique of the limits and failures of capitalism, which has 'accorded workers a leading role in effecting change, with an affirmatory emphasis on the potentialities created by new forms of labour'. In this critique, precarity can be considered and mobilised as a 'political rallying point' (Gill and Pratt 2008: 20). I suggest the entrenched precarity of volunteer radio production can be usefully redeployed as such a rallying point, for pooling and sharing reflections on experiential commonalities and examples of good practice, and for developing new collective initiatives such as a volunteer bill of rights for radio practitioners.

CREATING ALTERNATIVE PUBLICS

In facilitating and sustaining complex spaces of community self-representation, I argue that community radio programmes enable the creation of alternative publics, which in turn facilitate articulation of shared experiences and strategies for negotiating the specificities of marginalised subjectivities. The capacities of these politicised alternative spaces have been considered in theoretical approaches as offering a discursive counterformation to that provided in dominant publics (the counterpublic) and an affective space of survival (an intimate public). In this section I want to consider ways in which the programmes discussed in this book variously embody and avail of the possibilities of these alternative discursive arenas. Nancy Fraser conceptualises the counterpublic as a structural response to exclusion experienced by members of marginalised groups, in recognition that 'members of subordinated social groups . . . have repeatedly found it advantageous to constitute alternative publics' (Fraser 1993: 14). Fraser argues for the importance of the counterpublic in its provision of an alternative space for political deliberation (following Jürgen Habermas) thus expanding discursive, as well as expressive and experiential space (Fraser 1993).[2] Catherine Squires builds on Fraser's analysis to consider the counterpublic further through its communicative and power relationships with a dominant public sphere (Squires 2002). For Squires, a counterpublic is typified by the following: ongoing communication between marginal and dominant public spheres, a reclamation of dominant and state-controlled public spaces, a strategic use of enclave spaces, and a remit to 'foster resistance, test arguments and strategies in wider publics; create alliances; persuade outsiders to change views; perform public resistance to oppressive laws and social codes; gain allies' (Squires 2002: 460). In this conceptual expansion of the counterpublic's capacities, Squires further elaborates its capacities for transcending a singular focus on identity affirmation by foregrounding interactivity in the context of power negotiations with the dominant public sphere: 'Differentiating the "dominant" public sphere from "counterpublics" solely on the basis of group identity tends to obscure other important issues, such as how constituents of these publics interact and intersect, or how politically successful certain publics are in relation to others' (Squires 2002: 447).

The functionality of such counter- and alternative publics is multifaceted and holistic. It expands beyond representational work where the political interests of a given dominated group can be adequately expressed and debated. Following this line of thought, Michael Warner underscores the functionality of alternative publics beyond representation:

> A public, or counterpublic, can do more than represent the interests of gendered or sexualised persons in a public sphere. It can mediate the most private and intimate meanings of gender and sexuality. It can work to elaborate new worlds of culture and social relations in which gender and sexuality can be lived, including forms of intimate association, vocabularies of affect, styles of embodiment, erotic practices, and relations of care and pedagogy. It can therefore make possible new forms of gendered or sexual citizenship—meaning active participation in collective world making through publics of sex and gender. (Warner 2002: 57)

These elaborations of meanings, relations and spaces are crucial for the purposes of listener identification with community-specific modes of being in the world, including access to community-specific practices and vernaculars. Lauren Berlant also complicates understandings of the functionality of the counterpublic, considering it alongside what she conceives of as an intimate public. Berlant differentiates between a political imperative inherent to the counterpublic and the affective capacities of the intimate public. From this standpoint, the counterpublic's functionality can be considered as primarily political while the intimate public's function is the community's affective survival (Berlant 2008). For Berlant, 'an intimate public is a space of mediation in which the personal is refracted through the general'; a collectively produced affective space of emotional contact. By identifying this function as a produced yet embodied way of 'being in the world', she proposes that

> an intimate public is an achievement. . . . Whether linked to women or other nondominant people, it flourishes as a porous, affective scene of identification among strangers that promises a certain experience of belonging and provides a complex of consolation, confirmation, discipline, and discussion about how to live as an *x*. (Berlant 2008: viii)

In her deliberations of the expressive nuances made possible within an intimate public, Berlant persistently recognises the ongoing messiness of belongingness, contradictions and slippages within it—as do Harvey and Joseph. She reminds us that publics are not neatly demarcated, their members clearly identifiable. As with conceptions of identity themselves, everyday experiences of a public—belonging to it, negotiating its boundaries and codes of practice, defining it to others—are messy and inconsistent. This may be why publics created through media forms attain and retain their cultural currency, even as they are dipped into and stepped away from.

Through its capacity for reinforcing familiarity through regular daily or weekly broadcasts, radio is well situated to produce and reinforce such fluid practices and spaces. Community radio's particular imperative of community-led representation means that these spaces are de facto at the service of a given community for deployment as a counterpublic or an intimate public. This enhanced imperative can be further understood by recognising radio's capacity to create a shared public space through the intimacy produced by liveness, as Alfred L. Martin argues: '[t]hrough radio's liveness, the host and listeners, while physically distant, co-create a synchronicity that engenders notions of community. . . . Such liveness mimics a virtual public square that encourages listeners' "active" listening and participation among a specialised group of listeners' (Martin 2018: 250). Thus, a community-produced programme can enable an alternative space fostering inclusivity and interactivity among community members. As Kate Lacey argues, '[a]ny intervention in the public sphere is always taken in the hope, faith or expectation that there is a public out there, ready to listen and to engage' (Lacey 2013: 7).

Recognition of the complexity, messiness and diverse capacities for representation made possible through the production and reinforcing of alternative publics allows for deeper analysis of modes through which practitioners navigate and reinforce their given communities. Intimate publics in particular can create the conditions for production of such structures of feeling. These structures of feeling can enable affective survival through shared recognition of articulated experiences. I suggest in chapter 4 that the alternative space produced and sustained by the LGBTQ programme *Rainbow Country* created space for articulation of the impact of the Orlando shootings on Toronto's LGBTQ communities. Mark Tara's commitment to broadcasting individual experiences exemplifies an intimate public where the personal can indeed become articulated as the political. The acuity of Tara's response exemplifies the cultural work of the intimate public of *Rainbow Country*, embodying Berlant's equally acute summary of the necessity for intimate publics: 'In an intimate public one senses that matters of survival are at stake and that collective mediation through narration and audition might provide some routes out of the impasse and the struggle of the present, or at least some sense that there would be recognition were the participants in the room together' (Berlant 2011: 226). Berlant here connects collective work that foregrounds affective survival of the intimate public with the ways in which this work is performed, especially insofar as it connects participants who are likely remote from each other. Producing such structures of feeling is one mode through which the shared sense of community can be reinforced and rearticulated. In this sense, collective and self-aware recognition of what is expressed in such structures of feeling functions as a community-building process in David Harvey's formulation (Harvey 2001).

Given the complexities of expression and access, and ongoing negotiations of power relations, I suggest the programmes in this book embody characteristics of the counterpublic and the intimate public in ways which are representative of their constituent communities. Through its use of Haitian Creole, *The Morning Drive* engages with listeners at the level of linguistic intelligibility, as the presenters' mix of languages ensures that primarily Haitian speakers can follow the show's discussion along with English speakers. At the same time, the heard accents of Haitian Creole signify the show's target community. The familiarity of a particular accent functions to reinforce a sense of community independently to what is being said, as members of Miami's Haitian American communities can hear people on the radio speaking like themselves. A further layer of aesthetic, but also affective, community-building is located in the practitioners' choice to combine Creole and English, even though it is an approach that has been criticised in their community for being too informal. Through combining the two languages, the three practitioners put their own distinctive stamp on the show, and in the process they actively signify their own positionality. Thus *The Morning Drive* signals its affiliations by creating an intimate public through the affective appeal of the practitioners' accents and also embodying the functions of the counterpublic in its discussion of community-specific issues such as health and event information. Although representing a New Orleanian experience of Haitian community, WRBH's show *La Voix d'Haiti* also reaches out to and reinforces an intimate public of shared experience, but does so through broadcasting almost entirely in Haitian Creole. WRBH's *The Vietnamese Show* utilises a similar strategy in its emphasis on Vietnamese language content. *Desi Beats* at CKUT draws in South Asian listeners through curating and sharing culturally specific content of interest to the community in Montreal alongside interviews with South Asian community members. Yet producer-presenter Ayesha Baig additionally aims to bring in new listeners from Montreal, a goal reflected in the programme title which is both specific and accessible. First language programming serves both an affective and a representative function, and thus embodies elements of an intimate public alongside a counterpublic. WWOZ's dedicated and local music programming reinforces an affective community comprising fans of New Orleanian jazz and other regional musics featured on the station, such as Cajun and gospel, but also reaches out to those tuning in to access a sense of a New Orleans everyday cultural imaginary.

CKUT programme *Under the Olive Tree* engages with its listening public through the lingua franca of activism, in particular via discussion and coverage of the specificities of the ongoing Palestinian occupation. In her recognition that people come to the programme through their affiliation with the Palestinian struggle and therefore may have incomplete knowledge, programme producer-presenter Sawssan Kaddoura foregrounds accessible provision of information and interviews. In this politically necessary work, *Under the Olive Tree* embodies the functions of a counterpublic as conceived of in Squires's working definition—in particular through its strategies of fostering resistance and reinforcing alliances through the programme's ongoing work to inform new listeners and empower them to update their views. With *Rainbow Country* on CIUT, Mark Tara encourages newcomers to the programme and to the LGBTQ community to share and explain their positionality. In his discussion of queer culture and the necessity for a queer counterpublic, Warner identifies what he terms as the 'wrenching sense of recontextualization' experienced and negotiated by many in LGBTQ communities (Warner 2002: 198, 195). Tara's emphasis on providing a space for articulation of individual experiences facilitates and reinforces shared in participation and negotiation of LGBTQ experiences, thus functioning as an intimate public for listeners and contributors. Yet *Rainbow Country* also embodies the functionality of the counterpublic, through community representation at and coverage of events such as the town hall meeting and expressions of community responses following the Orlando shootings.

As Squires and Warner explain, alternative publics incorporate a range of practices particular to the given community. These practices can produce an active critique in the form of what Barnor Hesse terms transruptions. Transruptions, as conceptualised by Hesse, function as 'interrogative phenomena that, although related to what is represented as marginal or incidental or insignificant, that is identifiable discrepancies, nevertheless refuse to be repressed' (Hesse 2000: 17). It is the 'irrepressible quality of interrogation' which renders transruptions effective; as Hesse argues '[t]ransruptions are troubling and unsettling because any acknowledgement of their incidence or significance within a discourse threatens the coherency or validity of that discourse, its concepts or social practices' (Hesse 2000: 19, 17). In this productively unsettling activity, transruptions function as an extension of the work of the alternative public, availing of the affordances of the alternative space already challenging normative and reductive framings of the marginalised group. Through its mobilisation of listener knowledge of experiences particular to LGBTQ subjectivity, *Rainbow Country* productively enables a transruptive impulse which in turn provides an alternative mode of negotiating a recontextualised everyday. In articulating aspects of the Palestinian struggle in accessible and usefully informative ways, *Under the Olive Tree* produces a transruptive effect through its wider interrogation of the mechanisms informing the struggle, which by extension promotes recognition of the importance of alliance politics.

Tribal radio programmes constitute and reinforce their target publics differently in terms of scope and scale. Whereas the alternative publics for programmes discussed above comprise a space of shared affective and/or political affiliation and experiences usually circulating as a public among other publics, tribal conceptual spaces are predicated on the connectedness of preexisting tribal relationships which themselves reach back to the tribal lands from which the station broadcasts. Yet as I explore in chapter 5, online streaming of tribal content has a particular resonance for tribal members who have migrated away from the reservation for education, family, work or other reasons. For relocated tribal members, tribal programmes comprise a remote public through recognizable music and/or through culturally specific delivery modes with emotional resonance such as the 'shout-out'.

Given that any public comprises multiplicities even within a specific self-identified community, the complex articulations and negotiations which ensue remain particular to the communities and groups involved. In an ongoing reciprocal set of practices, alternative publics retain and indeed reproduce the local and cultural specificities of a show's conventions, by means of discursive reference points and debates particular to the programme. While counterpublics and intimate publics may initially have been formed in relation to a dominant mainstream public sphere, prior conceptions of a singular dominant public sphere which edges out representational spaces for marginalised groups are attenuating as spaces continue to proliferate across online platforms. While a hierarchy can still be traced between normative, dominant and (often better resourced) communicative spaces, and spaces which are clearly community-specific, the circulation and visibility of alternative community spaces means that alternative publics can be more easily found by people who need them.

REPURPOSING POLICY THROUGH COMMUNITY PARTICIPATION

To further combine the informing ideas of this book in a concluding set of arguments, I want now to consider ways in which community participation can more meaningfully be encouraged and enabled at structural and systemic levels. Community-led radio production provides community members with the means of production, and community ownership of radio stations deepens this structural control. However, as has been seen, many community-owned stations struggle for resources and must therefore continuously negotiate ways to transform their conditions of precarity. More meaningful community participation at the level of structural intervention could deploy practitioner experience and expertise in decision-making processes that could more effectively determine practice guidelines, terms of reference and governance structures, in turn producing a more specifically informed discursive and legislative regulatory framework for community radio practice. Consequently, such enhanced and empowered participation has the potential to endow practitioners with greater agency as they put their expertise to work. In this section, I consider two collaborative initiatives that enable further development of the North American community radio project at conceptual and structural levels. Throughout this book I have identified diverse elements of cultural work enabled by community radio structures and programmes. However, this work cannot necessarily be forensically mapped onto a set of direct benefits; as I argue above, the community-building practices of alternative publics are often messy, contradictory and certainly multifaceted in their complexity. While the societal values of volunteer production can be mapped to a point, as I do above, this cultural labour is at the same time subject to intensely individualised factors such as the volunteer's personal resources, available time and other competing responsibilities. Given this, I argue that the following areas of community-building commonly performed by community radio programmes and stations could usefully be better recognised and realised in policy frameworks: cultural preservation, including storytelling; Emergency Response services; first language programming; information provision service; and practitioner empowerment. A recurring preoccupation for many community radio stations is the extent to which their outreach work incorporates areas of community care which officially or otherwise should be undertaken by the state. Given this ongoing commitment to expansions of such care, this work needs to be formally legitimated in policy so that these activities are substantively supported and actively encouraged.

Sharing examples and strategies of good practice within the under-resourced contexts of tribal community radio provides value the skills to be learned from other practitioners, and through the reflective processes which typify such comparative practice. In the overarching context of community radio licensing in both Canada and the United States, Indigenous and tribal stations represent a substantial proportion of this provision. One reason for this in the United States is that tribal stations in remote rural locations face less competition in applying for or retaining a LP license, as the frequency provision is less crowded in rural areas and there is more bandwidth. At time of writing, there were well over one hundred tribal stations broadcasting in the United States. As I discuss in chapter 5, many tribal communities experience and negotiate ongoing difficulties at the profound registers of education, lack of infrastructure, and services. Consequently, for many tribal communities there is an acute need for the supportive services provided by tribal stations and a requirement for more robust intervention at the level of policy and resources. Each May, Native Public Media, an organisation for tribal broadcasters founded in 2004, hosts the Native Broadcast Summit for tribal practitioners to collectively discuss and share experiences and examples of good practice and provide opportunities to address ongoing challenges faced by many tribal stations.[3] Native Public Media (or NPM) is an organisation dedicated to supporting tribal broadcasters in community-building in media form, 'through media access, control and ownership', with a focus as follows:

> Native Public Media (NPM) provides services that encourage the expansion and strengthening of Native media through platforms that are community-based, local, and democratic. NPM, as a national center, provides leadership, centralized resources, and strategic and coordinated approaches to successfully strengthen the Native Broadcast System. These services include broadcast licensing guidance, Federal Communications Commission (FCC) and Corporation for Public Broadcasting (CPB) compliance, station operations guidance, legal guidance, broadcast leadership training and education, and telecommunications and communications policymaking.[4]

In 2018, the Native Broadcast Summit foregrounded 'sharing on the topics of station compliance and enhanced understanding of the Community Service Grant, technology and media, and community engagement'.[5] Presentations, discussions and other events were organised around the themes of 'community engagement', 'station operation' and 'technology and media'—with topics including harassment prevention, interviewing and reporting skills, sharing content across stations, uses of online platforms and youth storytelling. The summit concluded with a 'community service grant clinic' providing suggestions, guidelines and support for this process. As became clear in my consideration of their overall aims and conference interventions, the Native Broadcast Summit represents a collective and collaborative mode of sharing experiences in directive ways that contributes to better community radio service and offers a model for future interventions.

Finally, following my suggestion above for developing modes of community practitioner participation at structural levels, I want to turn to a consideration of possibilities within broadcast policy frameworks in order to identify and explore a particular productive intervention which can serve as a constructive model for policy formation. In a nutshell, policy can be considered and defined as a set of regulations that govern broadcast station and production practices. Active community participation in these processes is meaningful and necessary, as community members need to be involved in decision-making which directly informs their production practices and available modes of representation. Greater policy recognition of community radio's cultural goods could therefore facilitate protection of the rights of volunteer workers and more directly legitimate the wider cultural work of community radio.

These areas of community broadcasting practice are provided for in culturally specific ways in a set of collectively determined radio proposals from Indigenous communities that have been developed in an ongoing Canadian project to address limitations in current broadcasting policy on Indigenous radio provision. In the spring of 2017, the 'Future of First Nations, Inuit, and Métis Broadcasting: Conversation and Convergence' project in Canada hosted workshops on or near five Indigenous territories to identify and discuss problems with Canada's existing Indigenous broadcast provision with Indigenous radio practitioners, Indigenous representatives and representatives from other Canadian radio sectors. The aim of this collaborative and multi-sited process was to develop a set of new guidelines and practices through collective discussion for submission to the Canadian Radio-television and Telecommunications Commission (CRTC) to support changes in the existing policy.[6] The workshops took place at Winnipeg, Manitoba; Iqaluit, Nunavut; Edmonton, Alberta; Homalco First Nation, British Columbia; and Halifax, Nova Scotia. A workshop was also scheduled to take place in Chisasibi, Quebec, but this was cancelled as the organisers received no submissions. In advance of each event, a statement was provided that acknowledged the visitors' presence on Indigenous lands or on unceded territories as appropriate. At the summative conference in Ottawa, there was a formal recognition that the event was taking place on unceded lands, stated as follows: 'In preparation for The Future of First Nations, Inuit, and Métis Broadcasting national conference in Ottawa, we acknowledge these events are proposed to take place on the traditional unceded territory of the Algonquin Anishnaabeg people'.[7]

Territorial acknowledgements in recognition of temporary uses of Indigenous lands or of unceded territories are regularly (although not universally) incorporated into public events in Canada which take place on these lands. This symbolic acknowledgement prefacing Indigenous activities specific to the territories in each workshop strengthens a sense of collaboration, cultural and political recognition, and partnership at the structural register. Collating and presenting outcomes from the consultative workshops on Indigenous territories, the Ottawa programme balanced Indigenous-led panels with papers from CRTC policymakers and academics and incorporated diverse forms of presentation and address to reflect the specificities of participating Indigenous peoples.

The extensive consultation process produced a draft 'Event Outcomes' document that summarised the 'key policy and process themes' from the five workshops, foregrounding input from Indigenous practitioners.[8] Arranged thematically, what emerged in these outcomes was a holistic representation of changes needed across consultation processes—including legal recognition and status, management and ownership, materialities, programming content, need for appropriate resources, and training and youth participation. These recommendations reflected the materialities of community radio practice in the discussion of Resources under Theme 1 of the 'Event Outcomes' document. This section called for 'the need for suitable and sustainable resources for Indigenous stations and broadcasters' with references to specificities of these resources. It further called for recognition of Indigenous radio stations as legal entities, with 'a fair system for allocating funding', alongside a further recommendation that neither policy frameworks nor funding mechanisms should operate under the assumption that Band Councils contribute to Indigenous stations ('Event Outcomes': 3). A recognition under Theme 1 that limits on advertising should be revisited reflected a common dilemma of community stations. For many stations and communities alike, broadcasting ads for local businesses benefits both, yet at the same time the number of small local businesses who can afford broadcast ads is often small enough to ensure a given station's airtime will not be dominated by such advertising. Theme 1's final point called for the requirement that a CRTC-issued mandate must be accompanied by funds to support required services such as the Emergency Broadcasting System.

The second and third themes synergistically foregrounded training and programming content. Theme 2 calls for local training and training in Indigenous languages, and Theme 3 provides details of preferred content quotas for Indigenous content (including 40 percent to comprise Indigenous language and 40 percent indigenous music) and a focus on Indigenous language in policy and the need for 'funding for Indigenous content (to sustain above quotas), particularly in Indigenous languages.' The following section, Theme 4, crucially identified the 'need to conduct *meaningful consultation* with Indigenous communities before, during, and after the policy review process' (emphasis in the original, 'Event Outcomes': 4). This section provides guidelines for meaningful consultation including the need for Indigenous representation in CRTC staff, the need to include face-to-face consultations with Indigenous committees, and the need for public consultations. An additional consideration called for conducting consultations in spaces which are 'welcoming to Indigenous broadcasters and participants' ('Event Outcomes': 4). A further recommendation called for the provision of 'food, not just snacks, as part of consultation activities'. This seemingly small detail demonstrates a substantive recognition of everyday materialities for many Indigenous practitioners, and also reflects my own observation at tribal station KWLP. For example, I joined a station meeting at KWLP which took place at 6 p.m.: since the meeting took about two hours and incorporated a peer training session, substantial food and bottled water were provided. Likewise, at the consultative Future of First Nations, Inuit, and Métis Broadcasting conference in Ottawa, ample and varied food was provided in a buffet at lunchtime, and throughout the day there were jokes about the less substantial 'snacks' doled out at other events outside the project.

Also under Theme 4's recommendations for meaningful consultation was a call for recognition of the 'need to accommodate oral traditions within consultation process to allow Indigenous people, especially elders, to say what they have to share and be heard without interruptions' ('Event Outcomes': 4). Inclusion of this recognition of the specificities of Indigenous modes of articulation and discussion engaged with ways in which the consultative process can itself be made more equitable and representative of Indigenous cultural norms and modes of address. The need for these specificities was reinforced in a subsequent section that provided guidelines for an 'Indigenous Journalism Manual of Best Practices' which foregrounds particularities of Indigenous content and modes of articulation. In their entirety, these read as follows:

a. Not cutting off an elder or questioning a person's truth.
b. Focusing on content that is positive, constructive, not airing programming that is triggering.
c. Focusing on sovereignty, nationhood, and the stewardship of land.
d. Climate change necessitates broadcasters sharing tools, like food sovereignty.
e. Need to address shame and language, among other critical under-covered issues.
f. Program structures include: self-locating or collective members stating on air where they from, and locating their place as broadcasters.
g. Gifts for guests, especially elders who have come on the program and helped to guide the show.
h. Making a welcoming studio space, goal is many voices.

i. Strive for self-representation by giving interview guests questions in advance or a chance to listen to an interview and provide comments before it goes on air. ('Event Outcomes': 7)

In their foregrounding of the particularities of Indigenous language use, which differs across Indigenous cultures and peoples (as with the Arizona tribes) and within presenter and guest situatedness, these proposals both reflect and reinforce such best practices. The acknowledgement and inclusion of material factors alongside consultative and representative components demonstrates the value of this inclusive consultative process, and reflects the contribution of detailed Indigenous knowledge of radio practices and problems experienced at Indigenous stations.[9] Further outcomes of the project included a presentation of outcomes to the CRTC research department and a proposed intervention for the upcoming Native Broadcasting Policy (CRTC 1990–1989) review. This project's facilitation of a comprehensive, regionally and culturally specific consultative process enabled a truly community-driven set of policy-facing outcomes that constituted a reflective and sustained collaboration across First Nations, Inuit and Métis peoples. In its commitment to a multi-sited consultative process, the 'Future of First Nations, Inuit, and Métis Broadcasting' project facilitated Indigenous input into policy structures for their own radio stations. Crucially, it further exemplifies a consultative process model for future policy formation involving Indigenous and other marginalised communities.

PRACTITIONER PARTICIPATION AND RESEARCH PRACTICE

Insights from community radio practitioners have centrally shaped this book. In seeking out the meaning that practitioners give to their work in community radio, I gathered in-depth insights on the mechanics of ongoing radio production. In the process, I learned new things about the radio show text as perceived from the inside of its workings. I discovered how practitioners negotiated the (frequently delicate) politics of localised representation, how they maintained ongoing on-air discretion and how they persuaded guests to be interviewed on their show. I found that many practitioners moved time and space to continue presenting and producing their show, sometimes travelling long distances to the station, allowing their commitment to the show to inform what employment they took up or establishing standing arrangements at work to keep their show timeslot free. I discovered the degree to which many practitioners considered their production and presenting work as integral to their identity. I learned that some stations enforce a music policy, sometimes due to an overarching cultural remit and in other cases to ensure songs played did not advocate violence in ways that could trigger vulnerable listeners. I was heartened by the value station managers conferred on their volunteer producer-presenters, their commitment to the shows, and the loyalty of their listener base. And as discussed already, I found commonalities of experience everywhere. Seeing these commonalities coalesce in a holistic manner, I perceived more clearly the value each practitioner placed on their radio work. The phrase 'a passion for radio' is often repeated (and forms the title of at least one academic study of radio). As a subjective feeling, it may be too laden with emotion to quantify. Yet 'passion' is repeatedly cited by practitioners as a reason for continuing to work in radio, and therefore I include mention of its recurrence here, outside of my analytical typology, as it constitutes its own considerable value but rightly resists academic interrogation.

When conceiving of a comparative analysis of representations of diversity in community radio, I decided early on that I needed to further delve into these practitioner perspectives to discover what was actually experienced on the ground. As my research developed and began to take form, it became clear that such deeply experienced and felt perspectives would be central to my analysis. I therefore argue all through this book for the importance of facilitating meaningful community participation at all registers of community radio's structures and practices so as to produce a robust and sustainable set of conditions which would more thoroughly support and bolster the under-resourced project of community radio. While the majority of the cultural work of community radio remains undervalued in fiscal and legislative registers, I hope that I have argued effectively here for a much-needed recognition that this systemic neglect needs radically to change.

NOTES

1. In using the term 'Indigenous' here I am guided by regular reference to and discussion of Indigeneity in the Canadian context by First Nations, Inuit and Métis practitioners and representatives as well as policymakers. While in the US, as noted in chapter 5, tribal names, 'American Indian' and 'Native American' are all in use, in Canada Indigenous is more widely used, by my observation and within documentation produced in consultation with First Nations, Inuit and Métis practitioners. Additionally, I am guided by Indigenous and local uses of the term 'territories' used where these are acknowledged as formally unceded by Indigenous peoples or otherwise ne-

gotiated with the state. I use 'Indigenous' with the capital letter, following prevalent observed, discursive and documented practice during my fieldwork.

2. The broader term 'counterpublic' was first coined by Rita Felski with reference to feminist publics (Squires 2002: 450). Nancy Fraser coined the term 'subaltern counterpublic' to reflect the minoritized positionality of its participants (Fraser 1993: 14).

3. Many thanks to Hector Youtsey for making me aware of the annual Native Public Media summit held each May.

4. From https://www.nativepublicmedia.org/about (accessed 24 August 2018).

5. From https://www.nativepublicmedia.org/native-broadcast-summit (accessed 24 August 2018).

6. See http://www.indigenousradio.ca/About.php (accessed 10 August 2018).

7. This statement is reproduced on the project's website here: http://www.indigenousradio.ca/ (accessed 10 August 2018).

8. 'Event Outcomes', retrieved from http://www.indigenousradio.ca/resources/Event_Outcomes_Ottawa.pdf (accessed 10 August 2018).

9. The Future of First Nations, Inuit, and Métis Broadcasting project team comprises academics from media and law fields and a representative of the Wawatay Native Communication Network. The team includes John Gagnon, CEO of the Wawatay Network; Geneviève Bonin, professor in communication; anthropologist Kathleen Buddle; Gretchen King, a postdoctoral researcher in communication; and Chris Albinati, a PhD candidate studying law. Albinati and King are cofounders of the Canadian Community Media Advocacy Centre. Outcomes were published online and the project team invited clarifications and further feedback.

Bibliography

Ali, C. and Conrad, D. 2015. 'A Community of Communities? Emerging Dynamics in the Community Media Paradigm', *Global Media and Communication*. Vol. 11 (1), 3–23.

Alfred, T. 2005. *Wasáse: Indigenous Pathways of Action and Freedom*. Peterborough, ON: Broadview Press.

Baker, H. A., Jr. 1995. 'Critical Memory and the Black Public Sphere' in the Black Public Sphere Collective (eds.), *The Black Public Sphere*. Chicago/London: University of Chicago Press.

Banks, M. 2017. *Creative Justice: Cultural Industries, Work and Inequality*. London: Rowman & Littlefield International.

Barnard, S. 2000. *Studying Radio*. New York: Hodder Arnold.

Berlant, L. 2008. *The Female Complaint: The Unfinished Business of Sentimentality in American Culture*. Durham, NC: Duke University Press.

Berlant, L. 2011. *Cruel Optimism*. Durham, NC: Duke University Press.

Bessire, L. and Fisher, D. 2013. 'The Anthropology of Radio Fields', *Annual Review of Anthropology*. Vol. 42, 363–78.

Born, G. 2005. 'On Musical Mediation: Ontology, Technology and Creativity', *Twentieth-century Music*. Vol. 2 (1), 7–36.

Born, G. 2012. 'Mediating the Public Sphere: Digitisation, Pluralism, and Communicative Democracy' in Emden, C. J. and Midgley, D. (eds.), *Beyond Habermas: Democracy, Knowledge, and the Public Sphere*. New York: Berghahn Books.

Bosch, T. 2014. 'Community Radio' in Wilkins, K., Tufte, T., and Obregon, R., *The Handbook of Development Communication and Social Change*. Chichester, West Sussex: Wiley & Sons.

Bourdieu, P. 2010. *Distinction*. London: Routledge.

Brecht, B. 1930. 'The Radio as an Apparatus of Communication', in Willett, J. (ed., trans.), *Brecht on Theatre*. New York: Hill and Wang.

Campion, M. J. 2005. *Look Who's Talking: Cultural Diversity, Public Service Broadcasting and the National Conversation*. Oxford.

Carpentier, N. 2008. *Translocalism, Community Media and the City*. Working Paper, Center for Media Sociology. Available at http://homepages.vub.ac.be/~ncarpent/workpap/cemeso-07-translocalism.pdf (accessed August 2015).

Cheney, J. 2005. 'Truth, Knowledge and the Wild World', *Ethics & the Environment*. Vol. 10 (2), 101–35.

CKUT. 2016. *The Blue Book*. Available at https://ckut.ca/sites/ckut.ca/files/the_blue_book_2016-17.pdf (accessed January 2019).

Cody, F. 2011. 'Publics and Politics', *Annual Review of Anthropology*. Vol. 40, 37–52.

Collins, P. and Bilge, S. 2016. *Intersectionality*. Cambridge: Polity.

Couldry, N. 2010. *Why Voice Matters: Culture and Politics after Neoliberalism*. London: Sage.

Dawson, M. C. 1995. 'A Black Counterpublic? Economic Earthquakes, Racial Agenda(s) and Black Politics' in the Black Public Sphere Collective (eds.), *The Black Public Sphere*. Chicago/London: University of Chicago Press.

Downing, J. 2001. *Radical Media*. London: Sage.

Downing, J. and Husband, C. 2005. *Representing 'Race': Racisms, Ethnicities and Media*. London: Sage.

Dunbar-Hester, C. 2008. 'Geeks, Meta-Geeks, and Gender Trouble: Activism, Identity, and Low-Power FM Radio', *Social Studies of Science*. Vol. 38 (201).

Dunbar-Hester, C. 2010. 'The History and Future of Hyper-Local Radio', *The Atlantic*, 5 October 2010.

Dunbar-Hester, C. 2013. 'What's Local? Localism as a Discursive Boundary Object in Low-Power Radio Policymaking', *Communication, Culture & Critique*. Vol. 6, 502–24.

Dunbar-Hester, C. 2014. 'Producing "Participation"? The Pleasures and Perils of Technical Engagement in Radio Activism', *Public Culture*. Vol. 26 (1), 25–50.

Durham Peters, J. 1999. *Speaking to the Air*. Durham, NC: Duke University Press.

Echert, J. M. 2015. 'Radio Revolution: The Local Community Radio Act's Expansion of Possibilities for Low-Power FM Stations', *Washington Journal of Law, Technology and Arts*. Vol. 10 (3), Winter.

Fanon, F. 1967. *Black Skin, White Masks*. New York: Grove Weidenfeld.

Fauteux, B. 2013. 'Beyond Campus Borders: Canadian Campus Radio and Community Representation on the FM Dial', *The Radio Journal*. Vol. 11 (2), 137–53.

Fisher, D. 2012. 'From the Studio to the Street: Producing the Voice in Indigenous Australia', in Bessire, L. and Fisher, D., *Radio Fields: Anthropology and Wireless Sound in the 21st Century*. New York: NYU Press.

Fisher, D. 2013. 'Intimacy and Self-Abstraction: Radio as New Media in Aboriginal Australia.' *Culture, Theory and Critique*. Vol. 54 (3), 372–93.

Florini, S. 2015. 'The Podcast "Chitlin' Circuit": Black Podcasters, Alternative Media, and Audio Enclaves', *Journal of Radio and Audio Media*. Vol. 22 (2), 209–19.

Foy et al./Radio Centre-Ville. 1992. 'Inventing and Experimenting', in Girard, B. *A Passion for Radio*. Montreal/New York: Black Rose Books, 49–58.

Fraser, N. 1993. 'Rethinking the Public Sphere: A Contribution to the Critique of Actually Existing Democracy' in Bruce Robbins (ed.), *The Phantom Public Sphere*. Minneapolis/London: University of Minnesota Press.

Fraser, N. 2008. *Scales of Justice: Reimagining Political Space in a Globalizing World*. Cambridge: Polity.

Gill, R. 2008. 'Culture and Subjectivity in Neoliberal and Postfeminist Times', *Subjectivity*. Vol. 25, 432–45. doi:10.1057/sub.2008.28.

Gill, R. and Pratt, A. C. 2008. 'In the Social Factory? Immaterial Labour, Precariousness and Cultural Work', *Theory, Culture & Society*. Vol. 25 (7–8), 1–30.

Girard, B. 1992. *A Passion for Radio*. Montreal/New York: Black Rose Books.

Girard, C. and Grenier, G. 2008. 'Insulating an Ideology: The Enclave Effect on South Florida's Cuban Americans'. *Hispanic Journal of Behavioral Sciences*. Vol. 30 (4), 530–43.

Gomberg, P. 2007. *How to Make Opportunity Equal: Race and Contributive Justice*. Malden, MA: Wiley-Blackwell.

Gordon, J. (ed.). 2009. *Notions of Community: A Collection of Community Media Debates and Dilemmas*. Oxford: Peter Lang, 109–26.

Gordon, A. and Newfield, C. 1996. *Mapping Multiculturalism*. Minneapolis/London: Minnesota University Press.

Habermas, J. 1989. *Structural Transformations in the Public Sphere: An Inquiry into a Category of Bourgeois Society*. Boston: MIT Press.

Hartley, J. 2000. 'Radiocracy', in Hopkinson, A. and Tacchi, J. (eds.), *Special Issue: Radiocracy—Radio, Democracy and Development. International Journal of Cultural Studies*. Vol. 3 (2), August.

Harvey, D. 2001. *Spaces of Capital: Towards a Critical Geography*. Edinburgh: Edinburgh University Press.

Hendy, D. 2000. *Radio in the Global Age*. Cambridge: Polity Press.

Herman, E. S. and McChesney, R. W. 1997. *The Global Media: The New Missionaries of Corporate Capitalism*. London/New York: Continuum.

Hesse, B. 2000. 'Introduction: Un/Settled Multiculturalisms', in Hesse, B. (ed.), *Un/Settled Multiculturalisms: Diasporas, Entanglements, Transruptions*. London: Zed Books.

Husband, C. 2005. 'Minority Ethnic Media as Communities of Practice: Professionalism and Identity Politics in Interaction', *Journal of Ethnic and Migration Studies*. Vol. 31 (3), 461–79.

Jakobsen, J. 2005. 'Sex + Freedom = Regulation: Why?', *Social Text*. Vol. 23 (3–4) (84–85), 285–308.

Joseph. M. 2002. *Against the Romance of Community*. Minneapolis: Minnesota University Press.

Keith, M. K. 1995. *Signals in the Air: Native Broadcasting in America*. Westport, CT: Praeger Books.

King, G. 2017. 'History of Struggle: The Global Story of Community Broadcasting Practices, or a Brief History of Community Radio', *Westminster Papers in Communication and Culture*, Vol. 12 (2), 18–36.

Kruse, H. 2003. *Site and Sound: Understanding Independent Music Scenes: Music*. New York: Peter Lang.

Lacey, K. 2013. *Listening Publics: The Politics and Experience of Listening in the Global Age*. Cambridge: Polity.

Langer, J. 2005. 'Multicultural Radio in the Global Era: The Canadian Broadcaster Perspective', *Australasian Canadian Studies*. Vol. 23 (2), 113–40.

Lewis, P. M. 2008. *Promoting Social Cohesion: The Role of Community Media*. Strasbourg: Council of Europe.

Lewis, P. and Booth, J. 1989. *The Invisible Medium: Public, Commercial and Community Radio*. London: Macmillan.

Martin, A. L. 2018. 'Queer Infrequencies: SiriusXF's OutQ and the Limits of Queer Listening Publics', *Feminist Media Studies*. Vol. 18 (2), 249–63.

Mitchell, J. 2002. 'Lead us not into Temptation: American Public Radio in a World of Possibilities' in Hilmes, M. and Loviglio, J., *The Radio Reader: Essays in the Cultural History of Radio*. London: Routledge.

Mitchell, J. 2016. *Wisconsin on the Air: 100 Years of Broadcasting in the State That Invented It*. Madison: State Historical Society of Wisconsin.

Mohanty. S. 2001. 'Can Our Values Be Objective? On Ethics, Aesthetics, and Progressive Politics', *New Literary History: Objectivity in Ethics, Politics, and Aesthetics*. Vol. 32 (4), 803–33.

Moylan, K. 2009. 'Towards Accented Radio: Migrant Produced Programming in Dublin', in Gordon, J. (ed.), *Notions of Community: A Collection of Community Media Debates and Dilemmas*. Oxford: Peter Lang, 109–26.

Moylan, K. 2013. *Broadcasting Diversity: Migrant Representation in Irish Radio*. Chicago: Intellect/University of Chicago.

Moylan, K. 2018a. 'Accented Radio in Miami and New Orleans', in Badenoch, A. and Follmer, G. (eds.) *Transnationalizing Radio Research: New Approaches to an Old Medium*. Bielefeld: Transcript.

Moylan, K. 2018b. 'Accented Radio: Articulations of British Caribbean Experience and Identity in UK Community Radio', *Global Media and Communication*, Summer 2018.

Naficy, H. 2001. *An Accented Cinema: Exilic and Diasporic Filmmaking*. Princeton and Oxford: Princeton University Press.

Salazar, J. and Cordova, A. 2008. 'Imperfect Media and the Poetics of Indigenous Video in Latin America', in Wilson, P. and Stewart, M (eds.), *Global Indigenous Media: Cultures, Poetics, and Politics*. Durham, NC: Duke University Press.

Scannell, P. 1991. 'The Relevance of Talk', in *Broadcast Talk*. London: Sage.

Scifo, S. 2008. 'Community Radio', *Westminster Papers in Communication and Culture*. Vol. 5 (1).

Squires, C. 2002. 'Rethinking the Public Sphere: An Alternative Vocabulary for Multiple Black Public Spheres', *Communication Theory*. Vol. 12 (4), 446–68.

Szwarc, J. 2018. *Indigenous Broadcasting and the CRTC: Lessons from the Licensing of Native Type B Radio*. Canadian Radio-television and Telecommunications Commission. Available at https://crtc.gc.ca/eng/acrtc/prx/2018szwarc.htm (accessed 21 August 2018).

Tacchi, J. 1999. 'Radio Texture: Between Self and Others', in Miller, D. (ed.), *Material Cultures: Why Some Things Matter*. London: Routledge.

Tacchi, J. 2015. 'Ethnographic Action Research: Media, Information and Communicative Ecologies for Development Initiatives' in Bradbury-Huang, H. (ed.), *The SAGE Handbook of Action Research*. Los Angeles: Sage.

Tuhiwai Smith, L. 1999. *Decolonising Methodologies: Research and Indigenous Peoples*. London: Zed Books.

UNESCO, with Fraser, C. and Estrada, S. Restrepo. 2001. *Community Radio Handbook*. Paris: UNESCO.

Vertovec, S. 2008. 'An Accented Radio: Fostering Cosmopolitanism Through Media in Berlin', in Rovisco, M. and Nowicka, M. (eds.), *Cosmopolitanism in Practice*. London: Routledge.

Walia, H. 2013. *Undoing Border Imperialism*. Oakland, CA/Edinburgh: AK Press/Institute for Anarchist Studies.

Wall, T. 2007. 'Finding an Alternative: Music Programming in US College Radio', *Radio Journal: International Studies in Broadcast & Audio Media*. Vol. 5 (1), 35–54 (20).

Warner, M. 2002. *Publics and Counterpublics*. New York: Zone Books.

Williams, R. 1977. *Marxism and Literature*. Oxford: Oxford University Press.

Winskell, K. and Enger, D. 2014. 'Storytelling for Social Change' in Wilkins, K., Tufte, T., and R. Obregon, *The Handbook of Development Communication and Social Change*. Chichester, West Sussex: Wiley & Sons.

Index

All references to radio programmes are italicized.

NOTE: Descriptions for American Indian and First Nations peoples and tribes are here taken from each peoples' own chosen designation, where possible to determine this (e.g., Hopi tribe but Navajo Nation).

www.ingramcontent.com/pod-product-compliance
Ingram Content Group UK Ltd.
Pitfield, Milton Keynes, MK11 3LW, UK
UKHW021906150926
475560UK00003B/209

9 781783 489329